Sustainable Civil Infrastructures

Sustainable Civil Infrastructures (SUCI) is a series of peer-reviewed books and proceedings based on the best studies on emerging research from all fields related to sustainable infrastructures and aiming at improving our well-being and day-to-day lives. The infrastructures we are building today will shape our lives tomorrow. The complex and diverse nature of the impacts due to weather extremes on transportation and civil infrastructures can be seen in our roadways, bridges, and buildings. Extreme summer temperatures, droughts, flash floods, and rising numbers of freeze-thaw cycles pose challenges for civil infrastructure and can endanger public safety. We constantly hear how civil infrastructures need constant attention, preservation, and upgrading. Such improvements and developments would obviously benefit from our desired book series that provide sustainable engineering materials and designs. The economic impact is huge and much research has been conducted worldwide. The future holds many opportunities, not only for researchers in a given country, but also for the worldwide field engineers who apply and implement these technologies. We believe that no approach can succeed if it does not unite the efforts of various engineering disciplines from all over the world under one umbrella to offer a beacon of modern solutions to the global infrastructure. Experts from the various engineering disciplines around the globe will participate in this series, including: Geotechnical, Geological, Geoscience, Petroleum, Structural, Transportation, Bridge, Infrastructure, Energy, Architectural, Chemical and Materials, and other related Engineering disciplines.

SUCI series is now indexed in SCOPUS

and EI Compendex.

More information about this series at http://www.springer.com/series/15140

Hany Shehata · Mona Badr
Editors

Advancements in Geotechnical Engineering

The Official 2020 Publications of the Soil-Structure Interaction Group in Egypt (SSIGE)

Springer

Editors
Hany Shehata
Soil-Structure Interaction Group in Egypt
Cairo, Egypt

Mona Badr
German University in Cairo (GUC)
Cairo, Egypt

ISSN 2366-3405 ISSN 2366-3413 (electronic)
Sustainable Civil Infrastructures
ISBN 978-3-030-62907-6 ISBN 978-3-030-62908-3 (eBook)
https://doi.org/10.1007/978-3-030-62908-3

This Springer imprint is published by the registered company Springer Nature Switzerland AG
The registered company address is: Gewerbestrasse 11, 6330 Cham, Switzerland

Contents

SSIGE Official Publications 2020: Part 1

Assignment of Groundwater Table in Liquefaction Analysis of Soils

Muhsiung Chang[1](✉), Mei-Shan Chan[1], Ren-Chung Huang[1], Togani C. Upomo[2], and Rini Kusumawardani[3]

[1] Department Civil and Construction Engineering, National Yunlin University of Science and Technology (YunTech), Yunlin, Taiwan
{changmh,m10716011}@yuntech.edu.tw, huangrope@gmail.com
[2] Graduate School of Science and Technology, National Yunlin University of Science and Technology (YunTech), Yunlin, Taiwan
d10810212@yuntech.edu.tw
[3] Department Civil Engineering, Universitas Negeri Semarang (UNNES), Central Java, Indonesia
rini.kusumawardani@mail.unnes.ac.id

Abstract. This paper discusses the issue and potential influences of miss-assignment of groundwater table in the analysis of soil liquefaction. In viewing that the groundwater table (***GWT***$_0$) during subsurface exploration or testing for evaluating cyclic resistance ratio (***CRR***) of soils is sometimes mistakenly assumed the same as the groundwater table (***GWT***) for computing cyclic stress ratio (***CSR***) due to seismic shaking, the results of liquefaction analysis may thus be erroneous. If the ***GWT***$_0$ is assigned higher than the actual level, the ***CRR*** and the associated factor of safety ($\boldsymbol{F_L}$), would be overly predicted. Alternatively, if the ***GWT***$_0$ is assigned lower than its actual one, the ***CRR*** and $\boldsymbol{F_L}$ would be underestimated. If the groundwater table during exploration or testing is mistakenly assigned the same as the groundwater table for computing cyclic stress ratio (i.e., $\boldsymbol{GWT_0 = GWT}$; or "one-groundwater-table, ***OGT***, scenario"), then the variation in the groundwater tables will lead to the changes in ***CRR*** and ***CSR*** in the same sense. Owing to different rates of change, however, the computed factor of safety ($\boldsymbol{F_L = CRR/CSR}$), and the associated liquefaction potential index (***LPI***), may sometimes result in an unexpected situation. Namely, a rise in the groundwater tables would cause an unanticipated increase in the computed factor of safety and a decrease in the associated liquefaction potential index. Based on results of current study with assumption of ***OGT*** scenario, the ***LPI*** could be reduced by 10–30% if ***GWT***$_0$ is 3 m higher than the actual level; or alternatively, the ***LPI*** would be increased by 5–45% if ***GWT***$_0$ is 3 m lower than the actual one.

1 Introduction

Commonly adopted liquefaction analysis procedures, including methods based on $SPT - N$, $CPT - q_c$, and V_s, etc., involve separate evaluations of the cyclic resistance ratio (*CRR*) of soils and the cyclic stress ratio (*CSR*) due to seismic shaking (Youd

H. Shehata and M. Badr (Eds.): *Advancements in Geotechnical Engineering*, SUCI, pp. 3–18, 2021.
https://doi.org/10.1007/978-3-030-62908-3_1

et al. 2001; Seed et al. 1985; Robertson and Wride 1998; Andrus and Stokoe 2000). A factor of safety (F_L) against soil liquefaction at a particular depth of concern is obtained by the ratio of the above two evaluations, i.e., $F_L = CRR/CSR$. To assess liquefaction potential of a soil deposit, a depth-weighted approach is usually adopted with the consideration of liquefaction potential decreasing with the depth. In view of liquefaction incidents that had often observed within 20 m below the ground surface, Iwasaki et al. (1982) proposed a depth-weighted procedure, as shown below, by integrating the calculated factors of safety with depth, which has become most-widely used as the definition of liquefaction potential index (LPI or P_L):

$$\begin{aligned} LPI\ or\ P_L &= \int_0^{20m} F \cdot w(z)dz \\ \text{where } F &= 1 - F_L \qquad \ldots \text{ for } F_L \le 1.0 \\ F &= 0 \qquad \ldots \text{ for } F_L > 1.0 \\ w(z) &= 10 - 0.5z \end{aligned} \tag{1}$$

Iwasaki et al. (1982) further classified the assessed LPI or P_L into different categories: (1) $0 \le LPI \le 5$, for low liquefaction risk; (2) $5 < LPI \le 15$, for high liquefaction risk; and (3) $15 < LPI$, for very high liquefaction risk.

Figure 1 indicates a typical flowchart in conducting a liquefaction analysis. As can be seen, two groundwater data are required. The groundwater table, GWT_0, measured during the subsurface exploration and testing (e.g., SPT or CPT) is adopted for the assessment of CRR of soils, and the groundwater table, GWT, is assigned for the design earthquake condition and for the estimation of CSR due to seismic shaking. These groundwater tables are dissimilar and should not be confused, since they serve different purposes for the analysis and would have diverse effects on the assessment. Generally, the influences of GWT_0 and GWT on the computed factor of safety, F_L, and liquefaction potential index, LPI, would follow the trends as bellows (referring to Fig. 1 for symbols):

$$(GWT_0 \uparrow) \rightarrow (\sigma'_{v0} \downarrow) \rightarrow (N_1 \uparrow) \rightarrow (CRR \uparrow) \rightarrow (F_L \uparrow) \rightarrow (LPI \downarrow) \tag{2}$$

$$(GWT \uparrow) \rightarrow (\sigma'_{v} \downarrow) \rightarrow (CSR \uparrow) \rightarrow (F_L \downarrow) \rightarrow (LPI \uparrow) \tag{3}$$

It is clear that both groundwaters would influence the associated CRR and CSR in the same sense. One might expect an increase in the groundwater tables would soften the ground (or decrease effective stresses), thus reduce factor of safety against liquefaction and increase liquefaction potential. In view of the definition of factor of safety ($F_L = CRR/CSR$), however, a decrease in F_L by the increasing groundwater tables would not be certain, which would depend on the rates of change of GWT_0 and GWT on F_L. Supposed the rate of change of GWT_0 on F_L is more significant than that of GWT, then the same rise in both groundwater tables would lead to an increase in F_L, a situation appears contradictory with the intuition.

Whereas in general practices of liquefaction analysis, the groundwater table, GWT_0, during subsurface exploration and testing is often not employed for computing effective stresses σ'_{v0}, SPT blow counts N_1, and associated CRR. Instead, the groundwater table, GWT, assigned for design earthquake condition and for estimating cyclic stress ratio due

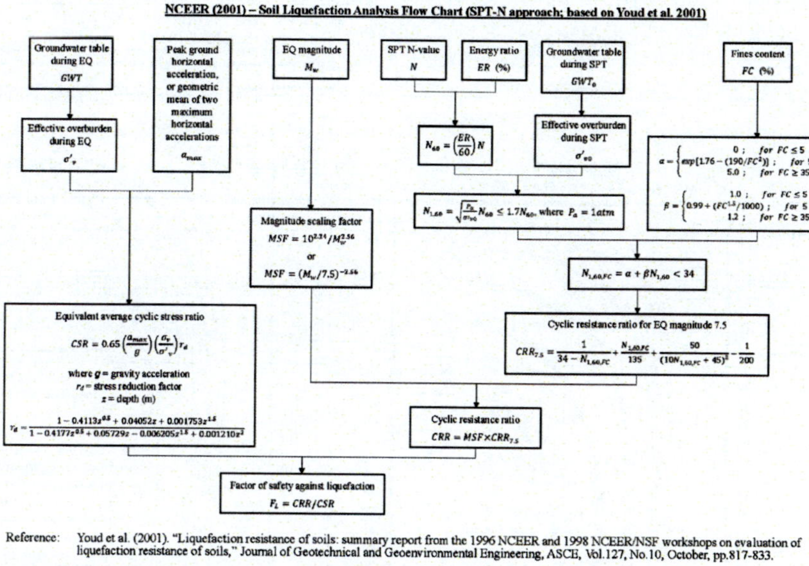

Fig. 1. Seed and NCEER SPT-N liquefaction analysis flowchart based on Youd et al. (2001)

to seismic shaking, is used in replacement of GWT_0 to compute σ'_{v0}, N_1 and CRR. The assignment of $GWT_0 = GWT$, termed hereafter as "one-groundwater-table" or OGT scenario, would be erroneous, since the CRR of soils is not properly evaluated by the groundwater table at the time of drilling or testing, and the incorrect GWT_0 might result in unanticipated F_L assessments as mentioned previously.

Figure 2 illustrates abnormal situations in the assessment of factor of safety due to OGT assumption. The factor of safety profiles are computed at various depths of a soil deposit with uniform $SPT - N$ distributions and different groundwater tables. As seen, the rise in the groundwater table would generally decrease the computed factor of safety. It is noticed, however, the rise in groundwater table could also lead to unexpected increases in the computed factor of safety, causing intersections of the factor of safety profiles as shown in the figure. These abnormal situations appear obvious for the cases with higher groundwater tables and shallower depths.

Figure 3 also indicates abnormal situations of the computed factor of safety, as dashed lines in the figure, due to the OGT assumption, where SPT blow counts are corrected with respect to GWT. Without OGT assumption, the GWT_0 would be irrelevant to GWT, and N_1 is solely determined by GWT_0, which is kept as a constant herein. Results demonstrate the computed factor of safety profiles without OGT assumption, as solid lines in the figure, are consistent as expected, where the rise in GWT will cause a decrease in F_L, and vice versa; with no intersection of the factor of safety profiles.

As discussed, the groundwater tables, GWT_0 and GWT, should not be misused in the liquefaction analysis of soils. Actually, this warning has been given by Youd et al. (2001) with the statements: "*The effective overburden pressure* σ'_{v0} *applied in (9;* $C_N = (P_a/\sigma'_{v0})^{0.5}$*) and (10;* $C_N = 2.2/(1.2+\sigma'_{v0}/P_a)$*) should be the overburden pressure at*

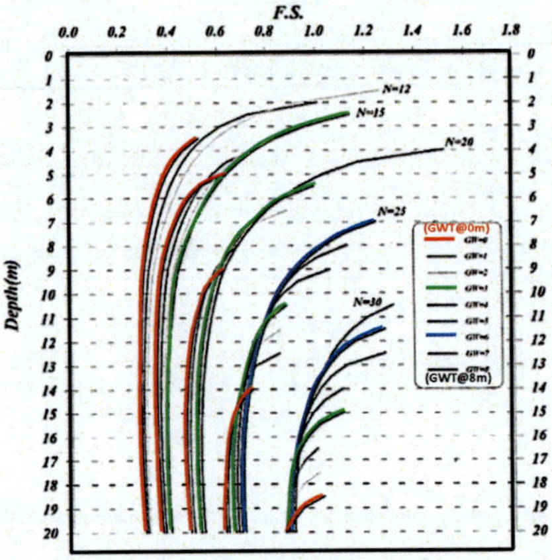

Fig. 2. Abnormal results of calculated factor of safety profiles due to one-groundwater-table assumption ($M = 7.5$, $PGA = 0.5g$, $FC = 10\%$, uniform $SPT - N$ deposits, Seed/NCEER2001 method) (Lin et al. 2001)

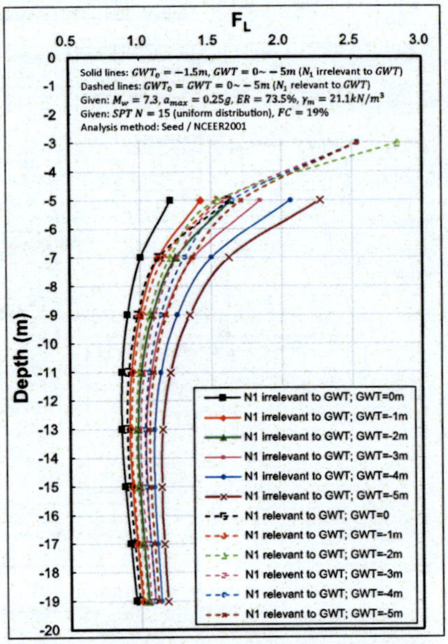

Fig. 3. Calculated factor of safety profiles due to OGT and non-OGT assumptions ($M = 7.3$, $PGA = 0.25g$, $FC = 19\%$, uniform deposit $SPT - N = 15$, Seed/NCEER2001 method) (Chang et al. 2011)

the time of drilling and testing. Although a higher groundwater level might be used for conservatism in the liquefaction resistance calculations, the C_N factor must be based on the stresses present at the time of the testing." (Note: C_N is a correction factor for $SPT - N$ values). In many occasions, however, the misuses of GWT_0 were found in routine practices of liquefaction analysis; part of reasons might have been due to the uncertainty or shortage of groundwater measurement data in the borehole logs that prevents proper assessments of cyclic resistance of soils (Wang et al. 2019).

The misuse of groundwater tables in liquefaction analysis has been addressed by Wang et al. (2019), based on the borehole data obtained from a recent liquefaction assessment project in Taipei basin. Preliminary results indicated the difference in the evaluated liquefaction potentials with or without OGT assumption would appear to be small. However, this finding was not conclusive since the groundwater tables, GWT_0, were measured during site explorations in flood seasons, which were generally close to the groundwater tables, GWT, assigned for the liquefaction analyses.

In accordance, the aims of this paper are to further address the issue of misuse of groundwater tables in liquefaction analysis of soils and to examine potential influences due to the misuse on the assessment of factor of safety and associated liquefaction potential. Several cases are employed for the examinations, including a fictitious borehole, actual boreholes, and a study site with an area of about 120 km^2 and a borehole number of more than 300.

2 Liquefaction Assessments of a Fictitious Borehole

A fictitious borehole is adopted to examine the influence of *OGT* assumption on the computed factor of safety and liquefaction potential. The borehole is assumed in a uniform stratum with a constant $SPT - N$ value of 10 and a fines content of 20%. The design earthquake is assumed with $M_w = 7.3$ and $a_{max} = 0.25g$. Seed/NCEER $SPT - N$ based method (Youd et al. 2001) is adopted for liquefaction analysis of soils at various depths and the depth-weighted procedure by Iwasaki et al. (1982) is employed for assessment of liquefaction potential in soil deposit up to 20 m deep.

Results of *OGT* scenario are compared with those of non-*OGT* scenario, in which the groundwater table, *GWT*, for estimating cyclic stress ratio due to seismic shaking is assumed varied from the ground surface to a depth of –5 m. Considering the groundwater table, GWT_0, for assessment of cyclic resistance of soils that could be higher or lower than its actual level of –2.5 m, the comparisons are discussed separately with results shown in Figs. 4 and 5, as well as in Tables 1 and 2, respectively.

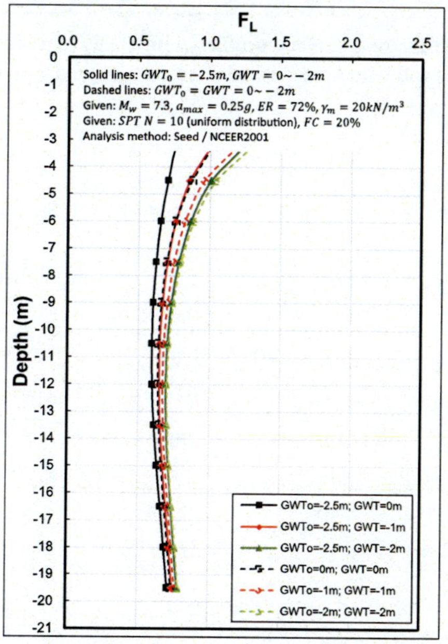

Fig. 4. Calculated factor of safety profiles with *OGT* (dashed lines) and without *OGT* (solid lines) assumptions and where the dashed lines are assigned with GWT_0 higher than its actual level of –2.5 m

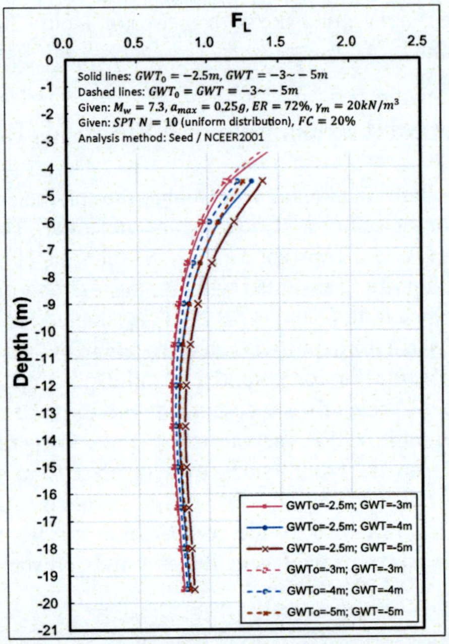

Fig. 5. Calculated factor of safety profiles with *OGT* (dashed lines) and without *OGT* (solid lines) assumptions and where the dashed lines are assigned with GWT_0 lower than its actual level of –2.5 m

Table 1. Calculated liquefaction potential indices with *OGT* (case I-A) and without *OGT* (case II-A) assumptions and where case I-A is assigned with GWT_0 higher than its actual level of –2.5 m

GWT	*LPI* or *PL* (%)		Error in *LPI* computations for case I-A
	Case I-A	Case II-A	
	$GWT_0 = GWT$	$GWT_0 =$ −2.5 m	
	$GWT = 0 \sim$ −2 m	$GWT = 0 \sim$ −2 m	
0	18.46	26.46	−30.2%
−1 m	15.19	18.37	−17.3%
−2 m	12.43	13.11	-5.1%

Figure 4 and Table 1 indicate the results of *OGT* scenario (dashed lines; Case I-A) and the associated non-*OGT* scenario (solid lines; Case II-A), where the groundwater table, GWT_0, in the *OGT* scenario is higher than its actual level of –2.5 m. It is noticed when GWT_0 is mistakenly assigned the same as *GWT* and also higher than its actual

Table 2. Calculated liquefaction potential indices with *OGT* (case I-B) and without *OGT* (case II-B) assumptions and where case I-B is assigned with GWT_0 lower than its actual level of –2.5 m

GWT	*LPI* or *PL* (%)		Error in *LPI* computations for case I-B
	Case I-B	Case II-B	
	$GWT_0 = GWT$	$GWT_0 = -2.5\,\text{m}$	
	$GWT = -3 \sim -5\,\text{m}$	$GWT = -3 \sim -5\,\text{m}$	
−3 m	10.18	9.50	7.2%
−4 m	8.10	6.73	20.4%
−5 m	6.54	4.48	46.2%

level (i.e., *OGT* scenario; dashed lines; Case I-A), the *CRR* of soils would be overestimated, leading to a higher estimation of F_L and a lower estimation of *LPI*; a situation appears contradictory with the intuition. For the fictitious example discussed herein, the computed *LPI* would be decreasing (i.e., less conservative) by 5–30%, if GWT_0 is 3 m higher than the actual level.

Figure 5 and Table 2 indicate the results of *OGT* scenario (dashed lines; Case I-B) and the associated non-*OGT* scenario (solid lines; Case II-B), where the groundwater table, GWT_0, in the *OGT* scenario is lower than its actual level of –2.5 m. It is noticed when GWT_0 is mistakenly assigned the same as *GWT* and also lower than its actual level (i.e., *OGT* scenario; dashed lines; Case I-B), the *CRR* of soils would be underestimated, leading to a lower estimation of F_L and a higher estimation of *LPI*; a situation appears contradictory with the intuition. For the fictitious example discussed herein, the computed *LPI* would be increasing (i.e., more conservative) by 5–45%, if GWT_0 is 3 m lower than the actual level.

The above discussions indicate the *OGT* assumption would be erroneous in liquefaction assessment of soils, leading to estimated factors of safety or liquefaction potentials contradictory with the intuitions. The fictitious example shows the liquefaction potential would be underestimated by 5–30%, if *OGT* is assumed and GWT_0 is higher than the actual level by 3 m. Alternatively, the liquefaction potential would be overestimated by 5–45%, if *OGT* is assumed and GWT_0 is lower than the actual level by 3 m.

3 Liquefaction Assessments of Actual Boreholes

The influences of *OGT* assumption on the computed factor of safety and liquefaction potential are further examined through actual boreholes drilled as a part of a field investigation program for the liquefaction study project in Huwei and Tuku Townships of Yunlin County, Taiwan, in 2019 (REI 2019). Two boreholes are selected from clayey or sandy deposits, and results of examination are discussed in the following subsections.

3.1 A Borehole in a Clayey Deposit

Borehole X6-1 is located in Huwei Township of Yunlin County, Taiwan. Table 3 indicates the material data, soil classifications and *SPT* blow counts within 20 m deep of the hole.

As seen, the borehole is in a clayey-silty deposit with relatively few sandy layers. Soil materials of the deposit are generally soft with $SPT - N$ values less than 10. The groundwater level GWT_0 measured during subsurface exploration is at –2.80 m.

Table 3. Material stratification and data at borehole No. X 6–1, in a clayey deposit of Huwei township, Yunlin county, Taiwan

Depth (m)	SPT N value	γ_m (kN/m^3)	FC (%)	LL (%)	PI (%)	USCS
−1.5	5	18.4	92	29	10	CL
−3.0	4	18.6	96	28	8	CL
−4.5	3	18.3	92	31	12	CL
−6.0	10	18.7	48	–	NP	SM
−7.5	5	18.3	48	–	NP	SM
−9.0	4	18.1	96	41	22	CL
−10.5	8	18.8	92	–	NP	ML
−12.0	6	18.2	93	42	19	CL
−13.5	5	18.5	95	–	NP	ML
−15.0	5	18.1	60	–	NP	ML
−16.5	10	18.6	34	–	NP	SM
−18.0	6	19.5	58	–	NP	ML
−19.5	16	18.2	31	–	NP	SM

Note: $GWT_{0,actual} = -2.80\,\text{m}$

A design earthquake with $M_w = 7.3$ and $a_{max} = 0.25g$ is assigned and a SPT hammer energy ratio $ER = 72\%$ is chosen. As before, Seed/NCEER $SPT - N$ based method (Youd et al. 2001) is adopted for the analysis and the depth-weighted procedure by Iwasaki et al. (1982) is employed for assessment of liquefaction potential. The examination on the influences of OGT assumption is performed with the analysis groundwater tables GWT varied from the ground surface to a depth of –6 m, which are approximately 3 m above or below the actual groundwater table during drilling at a depth of –2.80 m.

Results of the examination are shown in Table 4. For OGT scenario (Case I) and GWT_0 higher than its actual level, the computed liquefaction potentials will be underestimated. The computed LPI would decrease by about 20% if GWT_0 is 3 m higher than the actual level. Alternatively, for OGT scenario (Case I) and GWT_0 lower than its actual level, the computed liquefaction potentials will be overestimated. The computed LPI would increase by about 30% if GWT_0 is 3 m lower than its actual level. These findings are consistent with discussions in the previous section.

Table 4. Errors in liquefaction potential assessment due to one-groundwater-table assumption (case I) for borehole No. X 6–1, in a clayey deposit of Huwei township, Yunlin county, Taiwan

GWT	*LPI* or *PL* (%)		Error in *LPI* computations for case I
	Case I	Case II	
	$GWT_0 = GWT$	$GWT_0 = -2.8$ m	
	$GWT = 0 \sim -6$ m	$GWT = 0 \sim -6$ m	
0 m	13.80	16.97	−18.7%
−1 m	11.92	13.92	−14.4%
−2 m	10.70	11.04	−3.1%
−3 m	9.54	9.46	0.8%
−4 m	8.40	7.88	6.6%
−5 m	7.28	6.30	15.6%
−6 m	6.16	4.79	28.6%

Note: $GWT_{0,actual} = -2.80$ m
Analysis assumptions: $M_w = 7.3$, $a_{max} = 0.25g$, $ER = 72\%$
Analysis method: Seed/NCEER SPT-N-based method (Youd et al. 2001)

3.2 A Borehole in a Sandy Deposit

Borehole X2-2 is located in Tuku Township of Yunlin County, Taiwan. Table 5 indicates the material data, soil classifications and *SPT* blow counts within 20 m deep of the hole. As seen, the borehole is in a sandy deposit with relatively few silty layers. Soil materials of the deposit are generally loose to medium dense with $SPT - N$ values ranged about 5–15. The groundwater level GWT_0 measured during subsurface exploration is at –4.10 m.

A design earthquake with $M_w = 7.3$ and $a_{max} = 0.25g$ is assigned and a *SPT* hammer energy ratio $ER = 72\%$ is chosen. As before, Seed/NCEER $SPT - N$ based method (Youd et al. 2001) is adopted for the analysis and the depth-weighted procedure by Iwasaki et al. (1982) is employed for assessment of liquefaction potential. The examination on the influences of *OGT* assumption is performed with the analysis groundwater tables *GWT* varied from depths of –1 m to –7 m, which are approximately 3 m above or below the actual groundwater table during drilling at a depth of –4.10 m.

Results of the examination are shown in Table 6. For *OGT* scenario (Case I) and GWT_0 higher than its actual level, the computed liquefaction potentials will be underestimated. The computed *LPI* would decrease by about 10% if GWT_0 is 3 m higher than the actual level. Alternatively, for *OGT* scenario (Case I) and GWT_0 lower than its actual level, the computed liquefaction potentials will be overestimated. The computed *LPI* would increase by about 10% if GWT_0 is 3 m lower than its actual level. These findings are consistent with discussions in the previous section.

Table 5. Material stratification and data at borehole No. X 2–2, in a sandy deposit of Tuku township, Yunlin county, Taiwan

Depth (m)	SPT N value	γ_m (kN/m^3)	FC (%)	LL (%)	PI (%)	USCS
−1.5	6	15.6	32	–	NP	SM
−3.0	6	15.7	5	–	NP	SP-SM
−4.5	3	17.7	5	–	NP	SP-SM
−6.0	9	19.3	7	–	NP	SP-SM
−7.5	2.5	17.7	63	–	NP	ML
−9.0	9	18.1	76	30	6	ML
−10.5	1.5	16.7	97	32	4	ML
−12.0	2	18.4	72	–	NP	ML
−13.5	16	19.0	14	–	NP	SM
−15.0	15	19.5	20	–	NP	SM
−16.5	11	18.6	8	–	NP	SP-SM
−18.0	11	18.5	8	–	NP	SP-SM
−19.5	17	19.6	7	–	NP	SP-SM

Note: $GWT_{0,actual} = -4.10\,\text{m}$

4 Assessment of a Study Area with >300 Boreholes

The influences of OGT assumption on the computed factor of safety and liquefaction potential are also examined through a regional liquefaction study program carried in Huwei and Tuku Townships as mentioned previously (REI 2019). The study area, with about 120 km^2 in size and more than 300 boreholes, is chosen; in viewing that the examination could consider the variability of borehole data as well as the influence of OGT assumption on the distribution of liquefaction potentials at the site.

The total number of boreholes considered herein is 331, including 121 existing boreholes and 210 supplementary boreholes carried in the project (REI 2019). Based on current codes of Taiwan (MOI 2011 & 2017), a design earthquake is assigned with $M_w = 7.1$ at the project site and the peak ground acceleration is determined in accord with the type of soil deposit where the borehole is located, i.e., $a_{max} = 0.280g$ for Type III ground (soft deposits; $V_{s30} < 180m/s$) or $a_{max} = 0.308g$ for Type II ground (firm deposits; $V_{s30} = 180 \sim 270m/s$). It is noticed, however, most of the boreholes considered herein are situated in Type III ground, and only few are for Type II deposits.

The groundwater tables, $GWT_{0,proj}$, for assessing cyclic resistance of soils are based on the measurement during exploration and testing of the boreholes. Without a long-term monitoring data, the groundwater tables, GWT, for estimating cyclic stress due to shaking are assumed 3 m above or below the $GWT_{0,proj}$ at the borehole locations. In OGT scenario, the GWT_0 is assumed the same as GWT, and would not be the same

Table 6. Errors in liquefaction potential assessment due to one-groundwater-table assumption (case I) for borehole No. X 2–2, in a sandy deposit of Tuku township, Yunlin county, Taiwan

GWT	*LPI* or *PL* (%)		Error in *LPI* computations for case I
	Case I	Case II	
	$GWT_0 = GWT$	$GWT_0 = -4.1\,\text{m}$	
	$GWT = -1 \sim -7\,\text{m}$	$GWT = -1 \sim -7\,\text{m}$	
−1 m	38.33	42.89	−10.6%
−2 m	33.35	36.73	−9.2%
−3 m	29.22	30.57	−4.4%
−4 m	23.81	23.93	−0.5%
−5 m	15.19	14.42	5.3%
−6 m	13.42	12.27	9.4%
−7 m	10.76	10.21	5.4%

Note: $GWT_{0,actual} = -4.10\,\text{m}$

Analysis assumptions: $M_w = 7.3$, $a_{max} = 0.25g$, $ER = 72\%$
Analysis method: Seed/NCEER SPT-N-based method (Youd et al. 2001)

as $GWT_{0,proj}$. The energy ratio for *SPT* hammer is assumed as 72% for the existing boreholes. For supplementary boreholes, however, an energy ratio of 82% is assigned based on onsite calibrations for this project.

As before, Seed/NCEER $SPT - N$ based method (Youd et al. 2001) is adopted for liquefaction analysis and the depth-weighted procedure by Iwasaki et al. (1982) is employed for assessment of liquefaction potential. For liquefaction potential contour plotting, an interpolation technique by Inversed Distance Weighting, or IDW, is used.

Figures 6 and 7 show the results of *OGT* scenario (Case I-A) and the associated non-*OGT* scenario (Case II-A), where the groundwater table GWT_0 in the *OGT* scenario is higher than the project level $GWT_{0,proj}$ by 3 m. As shown in Fig. 6, when GWT_0 is mistakenly assigned the same as *GWT* and also higher than the project level (i.e., Case I-A), the computed *LPI* would be underestimated, with all the computation points falling on lower side of the diagonal line where *LPI* s are equal in Cases I-A and II-A. The average error in *LPI* computation of Case I-A is –17.3% (underestimated liquefaction potentials), for all the 331 boreholes evaluated.

Figure 7 shows distributions in area of computed liquefaction potentials for Cases I-A and II-A. The liquefaction potentials are divided into three levels, based on Iwasaki et al. (1982), where "$LPI \leq 5$" is for low liquefaction risk and shown in GREEN color, "$5 < LPI \leq 15$" is for high liquefaction risk and shown in YELLOW color, and "$15 < LPI$" is for very high liquefaction risk and shown in RED color. As seen, the computed liquefaction potentials for Case I-A are generally decreasing due to *OGT* assumption and GWT_0 higher than the project level. In terms of computed areas, the

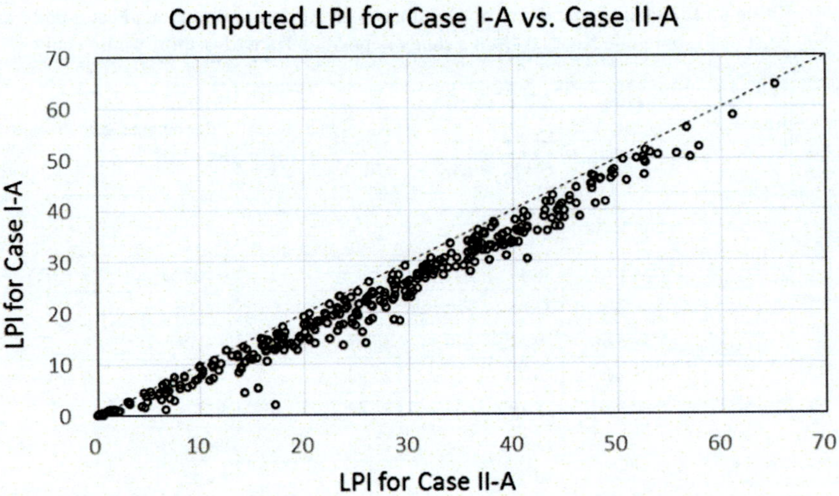

Fig. 6. Calculated indices of liquefaction potential with *OGT* (case I-A) and without *OGT* (case II-A) assumptions where the GWT_0 in case I-A is higher than its actual level by 3 m

Case I-A ($GWT_0 = GWT$)
($GWT_0 = GWT_{0,actual} + 3m$; $GWT = GWT_{0,actual} + 3m$)

Case II-A ($GWT_0 \neq GWT$)
($GWT_0 = GWT_{0,actual}$; $GWT = GWT_{0,actual} + 3m$)

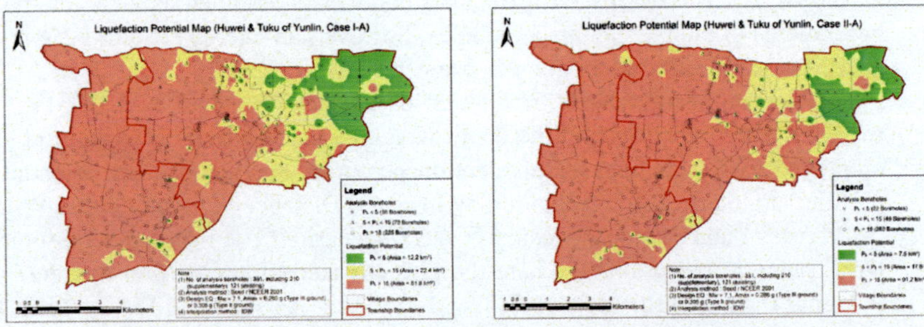

LPI Category	Computed Area in Case I-A (km²)	(%)
LPI ≤ 5	12.2	10.5
5 < LPI ≤ 15	22.4	19.2
15 < LPI	81.8	70.3
Sum:	116.4	100.0

Changes in Area

+4.1% ( )
+3.9% ( )
-8.0% ( )

LPI Category	Computed Area in Case II-A (km²)	(%)
LPI ≤ 5	7.5	6.4
5 < LPI ≤ 15	17.8	15.3
15 < LPI	91.2	78.3
Sum:	116.4	100.0

Fig. 7. Calculated areas of liquefaction potential category with *OGT* (case I-A) and without *OGT* (case II-A) assumptions where the GWT_0 in case I-A is higher than its actual level by 3 m

decrement in liquefaction potentials could be attributed to 8.0% reduction in RED zone, as well as 3.9% and 4.1% increments in YELLOW and GREEN zones, respectively.

Figures 8 and 9 show the results of *OGT* scenario (Case I-B) and the associated non-*OGT* scenario (Case II-B), where the groundwater table GWT_0 in the *OGT* scenario is lower than the project level $GWT_{0,proj}$ by 3 m. As shown in Fig. 8, when GWT_0 is mistakenly assigned the same as *GWT* and also lower than the project level (i.e., Case I-B), the computed *LPI* would be overestimated, with all the computation points falling on upper side of the diagonal line where *LPI* s are equal in Cases I-B and II-B. The average error in *LPI* computation of Case I-B is +24.3% (overestimated liquefaction potentials), for all the 331 boreholes evaluated.

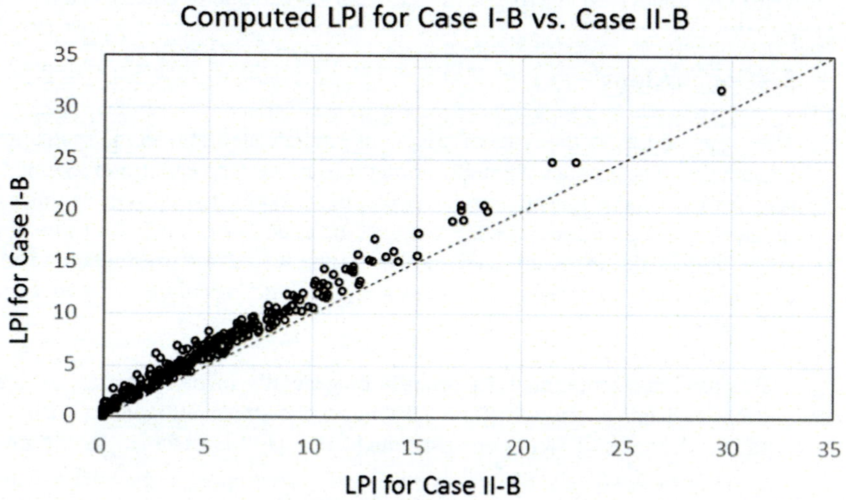

Fig. 8. Calculated indices of liquefaction potential with *OGT* (case I-B) and without *OGT* (Case II-B) assumptions where the GWT_0 in case I-B is lower than its actual level by 3 m

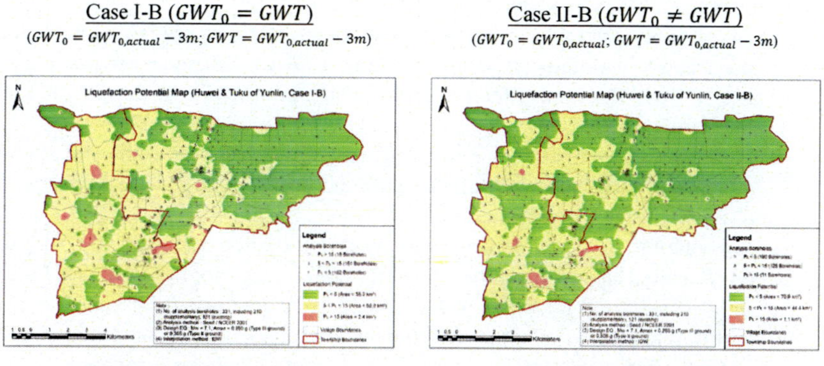

LPI Category	Computed Area in Case I-B	
	(km²)	(%)
LPI ≦ 5	55.2	47.4
5 < LPI ≦ 15	58.8	50.5
15 < LPI	2.4	2.1
Sum:	116.4	100.0

Changes in Area

⇐

-13.5% (green)

+12.4% (yellow)

+1.1% (red)

LPI Category	Computed Area in Case II-B	
	(km²)	(%)
LPI ≦ 5	70.9	60.9
5 < LPI ≦ 15	44.4	38.1
15 < LPI	1.1	1.0
Sum:	116.4	100.0

Fig. 9. Calculated areas of liquefaction potential category with *OGT* (case I-B) and without *OGT* (case II-B) assumptions where the GWT_0 in case I-B is lower than its actual level by 3 m

Figure 9 shows distributions in area of computed liquefaction potentials for Cases I-B and II-B. Similarly, the liquefaction potentials are divided into three levels, i.e., GREEN ($LPI \leq 5$), YELLOW ($5 < LPI \leq 15$), and RED ($15 < LPI$). As seen, the computed liquefaction potentials for Case I-B are generally increasing due to OGT assumption and GWT_0 lower than the project level. In terms of computed areas, the increment in liquefaction potentials could be attributed to 1.1% and 12.4% increases in RED and YELLOW zones, respectively, as well as 13.5% reduction in GREEN zone.

5 Conclusions

This paper addresses the issue of misuse of groundwater table in the evaluation of soil liquefaction. The groundwater table during subsurface exploration and testing, GWT_0, is used for assessment of cyclic resistance (or strength) of soils, but is often times assigned the same as the groundwater table assumed in the analysis, GWT, for estimating cyclic stress (or load) due to seismic shaking. The misuse would lead to erroneous results which contradict with the intuitions. Some key points and findings of the study are summarized as follows:

- Routine liquefaction analysis procedures generally involve separate evaluations of cyclic resistance ratio of soils (*CRR*) and cyclic stress ratio due to seismic shaking (*CSR*). A factor of safety against liquefaction (F_L) is obtained by the ratio of the above two, at a given depth of concern, and then integrated with depth to assess the liquefaction potential index (LPI) for the deposit.
- The groundwater table during exploration and testing (GWT_0) is used to calculate the effective stress and to assess the resistance or strength of soils, *CRR*, at the depth of concern. The increase in GWT_0 will decrease effective stress, which in turn will increase *CRR* and F_L, and thus decrease LPI.
- The groundwater table assumed in the analysis (GWT) is used to calculate the effective stress and to estimate the load of shaking, *CSR*, at the time earthquake. The increase in *GWT* will decrease effective stress, which in turn will increase *CSR*, decrease F_L, and thus increase LPI.
- These groundwater tables serve different functions in the analysis, so as to their influences on the assessed safety factor and liquefaction potential. Hence, GWT_0 and *GWT* should not be misused.
- General practices in liquefaction analysis would often times assign GWT_0 the same as GWT, called as one-groundwater-table (OGT) scenario; in which GWT_0, and *CRR* of soils as well, will go with *GWT* assumed in the analysis for estimation of *CSR*. Due to different rates of change of groundwater tables on the computed of safety factor and liquefaction potential, results of assessment would be erroneous.
- With OGT assumption and GWT_0 higher than actual level, the *CRR* of soils would be overestimated, thus leading to overestimated F_L and underestimated LPI. Alternatively, with OGT assumption and GWT_0 lower than actual level, the *CRR* of soils would be underestimated, and thus leading to underestimated F_L and overestimated LPI. Both of the above are erroneous and contradictory to intuitions.

- Based on results of current study, the *LPI* could be reduced by 10–30% for the case of *OGT* scenario and GWT_0 is 3 m higher than the actual level; or alternatively, the *LPI* could be increased by 5–45% for the case of *OGT* scenario and GWT_0 is 3 m lower than the actual one.

Acknowledgement. The authors would like to thank following peoples for discussions on the issue of one-groundwater-table in liquefaction analysis of soils: Professor T.S. Ueng of NTU, Taiwan, Professors C.H. Chen and J.H. Hwang of NCREE, Taiwan, Dr. C.C. Lu of NCREE, Taiwan, Mr. C.H. Wang of MAA, Taiwan, and the participants of 07/2018 NCREE seminar on the groundwater table issue for soil liquefaction. The opinions and insights provided through these discussions had been valuable and helpful in clarification of the issue that often times being neglected. The authors would also like to thank the Government of Yunlin County to provide borehole data for verification of the issue.

References

Andrus, R.D., Stokoe, K.H., II. Liquefaction resistance of soils from shear-wave velocity. J. Geotech. Geoenvir. Eng. ASCE **126**(11), 1015–1025 (2000)

Chang, M., Kuo, C.P., Shau, S.H., Hsu, R.E.: Comparison of SPT-N-based analysis methods in evaluation of liquefaction potential during the 1999 Chi-chi Earthquake in Taiwan. Comput. Geotech. **38**(3), 393–406 (2011)

Iwasaki, T., Tokida, K., Tatsuoka, F., Watanabe, S., Yasuda, S., Sato, H.: Microzonation for soil liquefaction potential using simplified methods. In: The 3rd International Earthquake Microzonation Conference, Seattle, vol. 3, pp. 1319–1330 (June–July 1982)

Lin, M.L., Chen, M.H., Shen, C.C.: Establishment of national liquefaction potential maps and assessment of liquefaction analysis methods in Taiwan. Report of NCREE, Taiwan, No. NCREE-01–17. (in Chinese with English Abstract) (2001)

Ministry of the Interior (MOI), Taiwan, Earthquake resistance design of structures – codes and explanations, by Construction and Planning Agency, Taiwan, promulgated on 19 Jan 2011. (in Chinese) (2011)

Ministry of the Interior (MOI), Taiwan. Design code for structure foundations (Draft), by Architecture and Building Research Institute, Taiwan, proposed in October 2017. (in Chinese) (2017)

Resources Engineering Inc. (REI). Report on field investigation and liquefaction potential assessment for enhancement of precision-scale of soil liquefaction potential maps. Yunlin County Mid-Precision Liquefaction Potential Maps and Liquefiable Ground Mitigation Demonstration Project (Phase I), for Yunlin County Government, Taiwan. (in Chinese) (2019)

Robertson, P.K., Wride, C.E.: Evaluating cyclic liquefaction potential using the cone penetration test. Can. Geotech. J. Ottawa **35**(3), 442–459 (1998)

Seed, H.B., Tokimatsu, K., Harder, L.F., Chung, R.M.: The influence of SPT procedures in soil liquefaction resistance evaluations. J. Geotech. Eng. ASCE **111**(12), 1425–1445 (1985)

Wang, J.S., Lu, C.C., Hwang, J.H., Huang, H.L.: The influence of the use of single ground water level on soil liquefaction assessment: a case study of Taipei basin. The 8th Japan-Taiwan Joint Workshop on Geotechnical Hazards from Large Earthquakes and Heavy Rainfalls, Kyoto, Japan, pp. 37–38 (October 2019)

Youd, T.L., Idriss, I.M., Andrus, R.D., Arange, I., Castro, G., Christian, J.T., Dobry, R., Liam Finn, W.D., Harder Jr., L.F., Hynes, M.E., Ishihara, K., Koester, J.P., Liao, S.S.C., Marcuson III, W.F., Martin, G.R., Mitchell, J.K., Moriwaki, Y., Power, M.S., Robertson, P.K., Seed, R.B., Stokoe II, K.H.: Liquefaction resistance of soils: summary report from the 1996 NCEER and 1998 NCEER/NSF workshops on evaluation of liquefaction resistance of soils. J. Geotech. Geoenvir. Engrg. ASCE **127**(10), 817–833 (2001)

Production Rate Estimation Using System Dynamics (Case Study: Clamshell Excavation at Top Down Construction)

Ibrahim Hamdan[1], Mohamed S. Eid[2(✉)], and Ahmed Elhakeem[2]

[1] Arab Academy for Science Technology, and Maritime Transport, Sheraton Heliopolis, Cairo 11799, Egypt
ihamdan_88@yahoo.com

[2] Construction and Building Engineering, Arab Academy for Science, Technology, and Maritime Transport, Sheraton Heliopolis, Cairo 11799, Egypt
{meid,aelhakeem}@aast.edu

Abstract. Construction projects usually face delays and do not finish on time due to many factors, including inaccurate estimation of production rates. Production rate for an activity depends on many factors that are interdependent and stochastic in nature. However, most of the available models for production rate estimation assume deterministic factors and seldomly account for the interdependencies between the various factors. This research aims to provide accurate estimation of production rates through accounting for stochastic and interdependent factors. To this effect, this paper presents the modeling and testing of a System Dynamics (SD) model for excavation activities using clamshell in top-down construction projects. SD allows for representing the interdependencies between the various elements and sub-processes of the construction process. In addition, the SD simulation process allows for depicting the stochastic nature of the various parameters that other estimation techniques fail to achieve. To achieve the desired outcome, the authors: (1) carried out a thorough literature review to study the factors affecting the excavation process; (2) collected data through questionnaires and site reports to quantify the impact of each factor on production rate and its occurrence likelihood; (3) modeled the excavation process through SD toolkit on AnyLogic simulation platform; and (4) tested and validated the model through the case study. The model was able to provide results within a mean square error of 4.14% throughout the testing phase. Although the excavation process of top-down construction activities has a lot of complexity, the presented SD model has proved able to simulate the process and provide desired realistic production rates. The model can be improved to account for more factors, in order to capture the full nature of the excavation activity, nevertheless, the modeling approach can be replicated to estimate the production rates for other complex construction activities.

1 Introduction

The quantity of work expected to be completed by a crew is during a given period of time is defined as "production rate". The production rate of an activity depends on the crew

H. Shehata and M. Badr (Eds.): *Advancements in Geotechnical Engineering*, SUCI, pp. 19–31, 2021.
https://doi.org/10.1007/978-3-030-62908-3_2

constituents, their interdependent processes, and various other stochastic parameters that affect the site operations. Thus, proper production rate estimation is a significant factor for project success through accurately estimating the project schedule. In addition, throughout the project, managers follow site production rate to determine if they need to increase productivity and revise the schedule (Gerald 2019). Also, many companies use production rate to create goals for their workers so that they have a working goal, or a target for the day, this helps to maintain and monitor workers to complete the project on time (Sarvaiya 2019).

Thus, there is a need for production rate estimation tool that allow project managers to accurately predict and estimate activities' duration based on stochastic site nature. Such tool should capture the actual size conditions based on project characteristics rather than relying only on historical productivity data of completed projects, or depending on the project manager's experience (Craig W. 1998).

2 Objective and Scope

The main goal of this research is to provide accurate estimation of production rates based on interactions and heterogeneity of crews. The research proposes a simulation model via System Dynamics (SD) to capture the stochastic variables to provide decision-makers with accurate production rate estimation.

The proposed model is developed and tested on clamshell excavation in top down construction. As a case study, the proposed model utilized actual gathered data from underground Metro of Cairo line3 phase 4A. The proposed model can be replicated and modified to address other cases and activities.

3 Background Information

3.1 Literature Review

In most construction companies, the production rates are usually established by handbooks that contain averaged production rates, historical data, and experts' opinions (Christian and Hachey 1995). It is difficult to accurately estimate the actual production rate based on those approaches as production rate is often significantly affected by other stochastic parameters (Ergan and Akinci 2012). Various methods were applied to overcome such problem utilizing Statistical and regression analyses, Expert System, Artificial Neural Network (ANN), and Simulation techniques.

The existing body of knowledge is rich with production rate estimation models based on statistical and regression analysis (Woldesenbet et al. 2012). This approach is considered a significant tool for data mining, dealing with big data analysis in a mathematical approach. AbouRizk and Halpin used a statistical approach to model the processes of construction activities (AbouRizk and Halpin 1992), while Chong et al. (2005) used regression analysis to predict production rate of foundations and the factors affecting it. Using t-test, a statistical comparison of man day values collected from firms were compared to state general unit price (Ulubeyli et al. 2014). Emad El-Maghraby (2014) built a regression model using correlation analysis of data collected from ten different construction sites to predict production rate of pouring ready mixed concrete activity.

Expert System approach is also used for production rate estimation. Expert systems are a system of if-then rules developed by field's experts to guide the decision-making and prediction of the process. Christian and Hachey (1995) studied the effects of delay times on production rates in construction, the breakdown of non-productive time enables the site management to focus onto the causes and factors of inefficient time and create efficient procedures to control operations. A prototype expert system was developed to obtain a much more accurate estimate of the probable production rate and provide a list of possible delays (Christian and Hachey 1995).

ANN approach is also used for production rate estimation. AbouRizk created a two-stage ANN model to predict labour productivity for industrial construction activities (AbouRizk et al. 2001). Another ANN model was developed that is able to learn from experience for improving the predicted labour productivity of concrete works at power plants construction (Heravi and Eslamdoost 2015).

Simulation modeling techniques are also used to predict production rates (AbouRizk 2010; Bokor et al. 2019). Simulation has different types; according to the data input, it may be stochastic and deterministic; and according to time as it could be continuous or discrete. Simulation models are useful to prototype, and replicate reality which can consider time change and deal with the dynamic and complex construction projects nature. Simulation models have three main types: Discrete-event Simulation models (DES), Agent base models (ABM), and continuous simulation models using system dynamics (SD).

DES is a process-oriented approach developed by Geoffrey Gordon 1960s. DES is based on the concept of entities and focuses on modeling the system in detail that assumes no change in the system between events, and jump from one event to another. Using DES, AbouRizk provided an overview of long-term simulation initiatives for the next generation of computer modeling system of construction (AbouRizk 2010).

ABM is a bottom up approach where the system focuses on the individual agents and their own behavior rules, who can learn, adapt, interact with each other and with the environment, and affect the system (Bokor et al. 2019). Using ABM, Watkins studied construction labour productivity. Production rate of an individual worker is called "skill level" which is a measure of number of work units that can be expected per unit time. The efficiency is the ratio of actual work and the work can be expected at their skill level. The model contains three labour crews and three tasks to complete to experiment the congestion, layoff crews and overstaffing principle (Watkins et al. 2009).

System Dynamics (SD) is a modeling technique to frame and analyze complex systems by modeling the different components that continually interact over time. The relationships and connections between model variables and components are the structure of the system. SD can provide a high-performance simulation by updating all variables concurrently with positive and negative feedbacks. SD enables for building simulations of complex systems to display the nonlinearity and long-term side effects of decisions (Sterman 2002).

A problem or a system may be represented a stock and flow diagram. A stock and flow model helps in studying and analyzing the system in a quantitative way. A stock is the term for any entity that accumulates or depletes over time, and flow is the rate of change in a stock (Grigoryev 2016). For example, the excavated quantities can be represented by stocks; Rates by flows; Feedback loops to describe routed systems feedback to itself.

The following is simple example of using SD in depiction an excavation activity in a construction project. Figure 1 represents a truck being loaded using an excavator with a certain loading rate. Upon loading, the truck moves to the dumping area to unload the excavated material. The simulation is affected by various parameters including the speed of the truck (Speed), distance travelled (Dist.), excavator loading rate (MachineLoadingRate), etc. Through accounting for the multiple factors and intercedences, the model can provide an estimate on the time required to finish the excavation process.

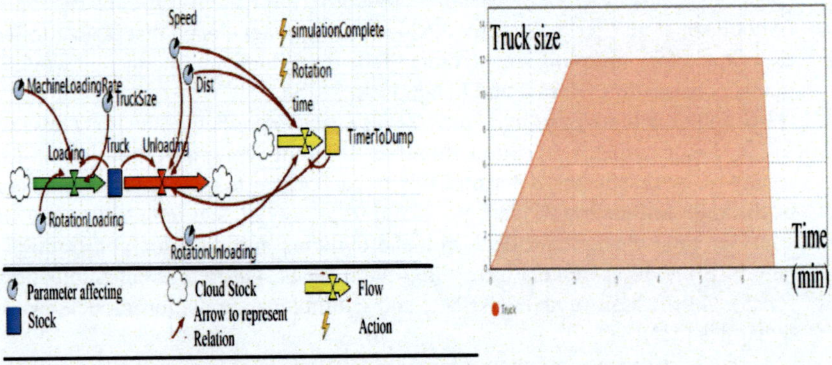

Fig. 1. Truck loading and unloading example of simulation modeling with AnyLogic

When simulating for a model; depending on the available data and the nature of the system, different problem may call for different methods. Complex problems cannot be described completely by using one basic simulation modeling approach; otherwise, a part of a problem may be outside the scope of the model or will represented less efficient and not matching the nature of the problem. Thus, the literature is also rich with hybrid models, where more than one simulation technique is utilized to address the problem and make use of the advantages of each approach. Using multi- method modeling can allow the best fitting method for each component of the model and combine all these components to form effective Hybrid simulation modeling (Borshchev and Filippov 2004).

Based on the current body of knowledge and the presented modeling approaches, the authors decided on utilizing SD as an overarching framework to capture the stochastic parameters and interdependences of the various factors of an activity. SD would allow for better evaluation and prediction of the project duration as it can accurately depict the various interdependent processes of the construction activities.

4 Methodology

To achieve the research objective, the following methodology have been developed: (1) review the existing body of knowledge regarding production rate studies, and literature for such operation history at the construction site to collect more data regarding the required crews and material for the excavation operation; (2) carryout surveys via questionnaires with experts (civil site engineers, QC engineers, consultant engineers, and general supervisors) who have more than three years of experience to narrow down and justify the factors and parameters affecting the project activities; (3) monitor and record site operation of the case study to collect data for each excavation box; (4) develop the model and integrate the data and parameters collected into the developed model using stocks, flows, parameters, etc.; (5) run the simulation model to get accurate production rate using Monte Carlo simulation and (6) test and validate this model by comparing the model results with the actual one from the field at each of the 11 excavation boxes.

5 Case Study

Excavation is the main work package on the critical path of constructing underground metro stations. Due to limited work-area available for constructing these stations, the construction process had to use a top down construction method that utilizes diaphragm walls. This method saves more storage area, enable better usage of cranes, and save scaffolding erection and striking. The clamshell's production rate at the Metro stations varies from excavation under slabs or at open cut and cover processes. This study focuses on modeling excavation activities under Metro Stations slabs via temporary openings at a top-down construction site. Field data for this research were collected from underground metro of Cairo line3 phase 4A at Heliopolis station. The main collected data are the factors affecting each excavation box, and the field daily excavation records for each excavation boxes, (this Model study excavation process of eleven different boxes).

5.1 Data Gathering Results

Based on the literature review and questioners distributed among experts to evaluate the relative importance of each factor, a pair-wise comparison matrix was developed. The matrix was calculated using the average values extracted from 15 respondents. The normalized comparison matrix and the criteria weights for each factor affecting are as per Table 1. The consistency check of the normalized matrix via consistency ratio (CR) is equal to 0.024. This is considered permissible as it is lower than 0.1. To this effect, matrix presented in Table 1 is considered consistent and reliable. The significant parameters found affecting the excavation are: (1) underground water, if it exists during excavation level, (2) clamshell cycle time; (3) truck cycle time with including capacity and availability; (4) production rate of mini-loaders; (5) safety regulations and rules; and (6) soil type.

The aforementioned factors can be either deterministic or stochastic. Thus, a questionnaire was performed to 10 field site experts depending on these parameters to determine which of these parameters are deterministic and which are probabilistic. The questionnaire result is as presented in Table 2

Table 1. Normalized comparison matrix.

	Weather effect	Soil type	Mini Loaders operator competent factor	Clamshell operator competent factor	Truck operator competent factor	Grading formation factor	Carpentry falls and removing for it	Lighting work rate	Preparation before start excavation factor	External Loading during night shift	Fueling factor	PR Mini Loader and enough no required	Swelling Factor	Bucket and truck full factor	Safety regulations and rules	Periodic maintenance of the equipments	under ground water if it exists	Waste time	Trucks effect	Clamshell cycle time factor	Criteria weight
Weather effect	[illegible]	[illegible]	[illegible]	[illegible]	[illegible]	[illegible]	[illegible]	[illegible]	[illegible]	[illegible]	[illegible]	[illegible]	[illegible]	[illegible]	[illegible]	[illegible]	[illegible]	[illegible]	[illegible]	[illegible]	[illegible]
Soil type	[illegible]	[illegible]	[illegible]	[illegible]	[illegible]	[illegible]	[illegible]	[illegible]	[illegible]	[illegible]	[illegible]	[illegible]	[illegible]	[illegible]	[illegible]	[illegible]	[illegible]	[illegible]	[illegible]	[illegible]	[illegible]
Mini Loaders operator competent factor	[illegible]	[illegible]	[illegible]	[illegible]	[illegible]	[illegible]	[illegible]	[illegible]	[illegible]	[illegible]	[illegible]	[illegible]	[illegible]	[illegible]	[illegible]	[illegible]	[illegible]	[illegible]	[illegible]	[illegible]	[illegible]
Clamshell operator competent factor	[illegible]	[illegible]	[illegible]	[illegible]	[illegible]	[illegible]	[illegible]	[illegible]	[illegible]	[illegible]	[illegible]	[illegible]	[illegible]	[illegible]	[illegible]	[illegible]	[illegible]	[illegible]	[illegible]	[illegible]	[illegible]
Truck operator competent factor	[illegible]	[illegible]	[illegible]	[illegible]	[illegible]	[illegible]	[illegible]	[illegible]	[illegible]	[illegible]	[illegible]	[illegible]	[illegible]	[illegible]	[illegible]	[illegible]	[illegible]	[illegible]	[illegible]	[illegible]	[illegible]
Grading formation factor	[illegible]	[illegible]	[illegible]	[illegible]	[illegible]	[illegible]	[illegible]	[illegible]	[illegible]	[illegible]	[illegible]	[illegible]	[illegible]	[illegible]	[illegible]	[illegible]	[illegible]	[illegible]	[illegible]	[illegible]	[illegible]
Carpentry falls and removing for it	[illegible]	[illegible]	[illegible]	[illegible]	[illegible]	[illegible]	[illegible]	[illegible]	[illegible]	[illegible]	[illegible]	[illegible]	[illegible]	[illegible]	[illegible]	[illegible]	[illegible]	[illegible]	[illegible]	[illegible]	[illegible]
Lighting work rate	[illegible]	[illegible]	[illegible]	[illegible]	[illegible]	[illegible]	[illegible]	[illegible]	[illegible]	[illegible]	[illegible]	[illegible]	[illegible]	[illegible]	[illegible]	[illegible]	[illegible]	[illegible]	[illegible]	[illegible]	[illegible]
Preparation before start excavation factor	[illegible]	[illegible]	[illegible]	[illegible]	[illegible]	[illegible]	[illegible]	[illegible]	[illegible]	[illegible]	[illegible]	[illegible]	[illegible]	[illegible]	[illegible]	[illegible]	[illegible]	[illegible]	[illegible]	[illegible]	[illegible]
External Loading during night shift	[illegible]	[illegible]	[illegible]	[illegible]	[illegible]	[illegible]	[illegible]	[illegible]	[illegible]	[illegible]	[illegible]	[illegible]	[illegible]	[illegible]	[illegible]	[illegible]	[illegible]	[illegible]	[illegible]	[illegible]	[illegible]
Fueling factor	[illegible]	[illegible]	[illegible]	[illegible]	[illegible]	[illegible]	[illegible]	[illegible]	[illegible]	[illegible]	[illegible]	[illegible]	[illegible]	[illegible]	[illegible]	[illegible]	[illegible]	[illegible]	[illegible]	[illegible]	[illegible]
PR Mini Loader and enough no required	[illegible]	[illegible]	[illegible]	[illegible]	[illegible]	[illegible]	[illegible]	[illegible]	[illegible]	[illegible]	[illegible]	[illegible]	[illegible]	[illegible]	[illegible]	[illegible]	[illegible]	[illegible]	[illegible]	[illegible]	[illegible]
Swelling Factor	[illegible]	[illegible]	[illegible]	[illegible]	[illegible]	[illegible]	[illegible]	[illegible]	[illegible]	[illegible]	[illegible]	[illegible]	[illegible]	[illegible]	[illegible]	[illegible]	[illegible]	[illegible]	[illegible]	[illegible]	[illegible]
Bucket and truck full factor	[illegible]	[illegible]	[illegible]	[illegible]	[illegible]	[illegible]	[illegible]	[illegible]	[illegible]	[illegible]	[illegible]	[illegible]	[illegible]	[illegible]	[illegible]	[illegible]	[illegible]	[illegible]	[illegible]	[illegible]	[illegible]
Safety regulations and rules	[illegible]	[illegible]	[illegible]	[illegible]	[illegible]	[illegible]	[illegible]	[illegible]	[illegible]	[illegible]	[illegible]	[illegible]	[illegible]	[illegible]	[illegible]	[illegible]	[illegible]	[illegible]	[illegible]	[illegible]	[illegible]
Periodic maintenance of the equipments	[illegible]	[illegible]	[illegible]	[illegible]	[illegible]	[illegible]	[illegible]	[illegible]	[illegible]	[illegible]	[illegible]	[illegible]	[illegible]	[illegible]	[illegible]	[illegible]	[illegible]	[illegible]	[illegible]	[illegible]	[illegible]
under ground water if it exists	[illegible]	[illegible]	[illegible]	[illegible]	[illegible]	[illegible]	[illegible]	[illegible]	[illegible]	[illegible]	[illegible]	[illegible]	[illegible]	[illegible]	[illegible]	[illegible]	[illegible]	[illegible]	[illegible]	[illegible]	[illegible]
Waste time	[illegible]	[illegible]	[illegible]	[illegible]	[illegible]	[illegible]	[illegible]	[illegible]	[illegible]	[illegible]	[illegible]	[illegible]	[illegible]	[illegible]	[illegible]	[illegible]	[illegible]	[illegible]	[illegible]	[illegible]	[illegible]
Trucks effect	[illegible]	[illegible]	[illegible]	[illegible]	[illegible]	[illegible]	[illegible]	[illegible]	[illegible]	[illegible]	[illegible]	[illegible]	[illegible]	[illegible]	[illegible]	[illegible]	[illegible]	[illegible]	[illegible]	[illegible]	[illegible]
Clamshell cycle time factor	[illegible]	[illegible]	[illegible]	[illegible]	[illegible]	[illegible]	[illegible]	[illegible]	[illegible]	[illegible]	[illegible]	[illegible]	[illegible]	[illegible]	[illegible]	[illegible]	[illegible]	[illegible]	[illegible]	[illegible]	[illegible]

6 Model Assumptions

To develop the model, the authors assumed that:

1. The excavation is at top bottom construction via temporary box opening only at Metro station;
2. The clamshell is used at excavation below slab;
3. The level of underground water table is below excavation level, and there are no injection pipes for the soil;

Table 2. Questionnaire results showing the type of parameters affecting Clamshell productivity.

Parameters / Variables affecting production rate	Sum of 10 respondents		Final (Probabilistic or Deterministic)
	D	P	
Number of Mini Loaders	10	0	D
Soil type	8	2	D
Mini Loaders operator competent factor day shift	9	1	D
Mini Loaders operator competent factor night shift	9	1	D
PR of One Mini Loader	1	9	P
Grading formation factor	2	8	P
Clamshell bucket size	10	0	D
Bucket speed up	8	2	D
Clamshell aim time box factor	1	9	P
Lag between trucks time	8	2	D
No of trucks	10	0	D
Carpentry falls time	1	9	P
Lighting work rate	8	2	D
Area of excavation	10	0	D
Excavation depth	10	0	D
Bucket speed down	8	2	D
Bucket pass from roof to top of excavation at start	10	0	D
Clamshell aim time truck and unloading the bucket	0	10	P
Clamshell Bucket collecting time	1	9	P
Clamshell competent operating factor day shift	9	1	D
Clamshell competent operating factor night shift	9	1	D
Distance from excavation location to dumping area	10	0	D
Truck Hauling speed	0	10	P
Truck dumping Time	2	8	P
Preparation time before start excavation	2	8	P
Falling carpenter depth	8	2	D
Distance from Dumping area to excavation	10	0	D
Truck Return speed	0	10	P
External Loading effect factor	7	3	D
Fuelling Time	2	8	P
Temp opening size factor	10	0	D
Truck Fulling Factor	0	10	P
Swelling Factor	9	1	D
Stock quantity of excavated material to be collected by Mini Loaders	9	1	D
Bucket full factor	0	10	P
Size of truck	10	0	D
Time Fuelling equipments	9	1	D
Time Collecting falling Formwork	2	8	P

4. Labor or operator fatigue is not considered through the model; and
5. Congestion between mini-Loaders effect is not considered while working at excavation and hauling material under the temporary box opening.

7 Model Development and Automation

The presented model is illustrated in Fig. 2. The model is mainly consisting of three main flows connecting four stocks to represent the main structure system of this model. The first flow is "*Mini Loaders*" that controls excavation process using mini-loaders in the excavation box. The quantity excavated by the mini-loaders is represented by stock "*Volume to be excavated*". The excavated material is collected under the temporary box opening, represented by a separate stock "*Stock of excavated material*". It should be noted that "*Volume to be excavated*" stock has an initial value depends on area of excavation, excavation depth, and the swelling factor; and can be defined according to Eq. (1).

$$\text{Initial value} = \text{area of excavation} * \text{excavation depth} * (1 + \text{swelling factor}) \quad (1)$$

"*Mini Loaders*" flow is mainly affected by its production rate "*PR mini loaders*" that depends on number of mini-Loaders, operator competency, the production rate of one mini-Loader, and grading for formation level effect (to adjust the excavation levels and the formation shape required specially if there are a lot of details).

The second flow "*Clamshell*" represents clamshell excavation process from the excavated material under the temporary opening "*Stock of excavated material*" to another stock "*Collected material by clamshell*". The "*Clamshell*" flow is mainly affected by its production rate "*PR clamshell*" that depends on the clamshell bucket size, operator competency, and the clamshell cycle time. Clamshell cycle time depends on bucket collecting time, speed up and down of the bucket, average height of excavation using clamshell, aim time box (to aim the bucket into the center of the opening before let it down), and aim time truck (to aim the bucket inside the dump box of tipper truck before unload it), plus bucket unloading time.

The third flow "*Unloading using trucks*" transfers the "*Collected material by clamshell*" to "*dumping area*" stock using a number of trucks. "*Unloading using trucks*" flow is mainly affected by the trucks production rates "*PR trucks*" that depends on size of truck, number of trucks, and truck total cycle time. Truck total cycle time depends on the distance from excavation box to dumping area, distance from the dumping area to the excavation box, speed of hauling, speed of return, dumping time, external loading effect at night shift, and lag between trucks.

It should be noted that some safety processes were simulated in the model. For example, the clamshell and mini-loaders cannot work together under the temporary opening, both working in rotation. While the clamshell is working, no mini loader is allowed to enter under the temporary opening but the mini loader can collect the excavated soil just before the opening, and while the mini loaders are pushing the collected excavated soil under the temporary opening of the box, the clamshell is not allowed to work.

The model was developed and automated using AnyLogic (version 8.4.0) (Borshchev 2013), using SD tools to represent the construction process. AnyLogic enabled adding all the data and parameters in the developed SD model. A simulation output using Monte Carlo technique for each excavation box (example of box's parameters data is presented in Table 3), automatically collects the time required to finish the excavation process at each time run - the desired accuracy of the result depends on number of iterations. Figure 3 presented a histogram of the model's results

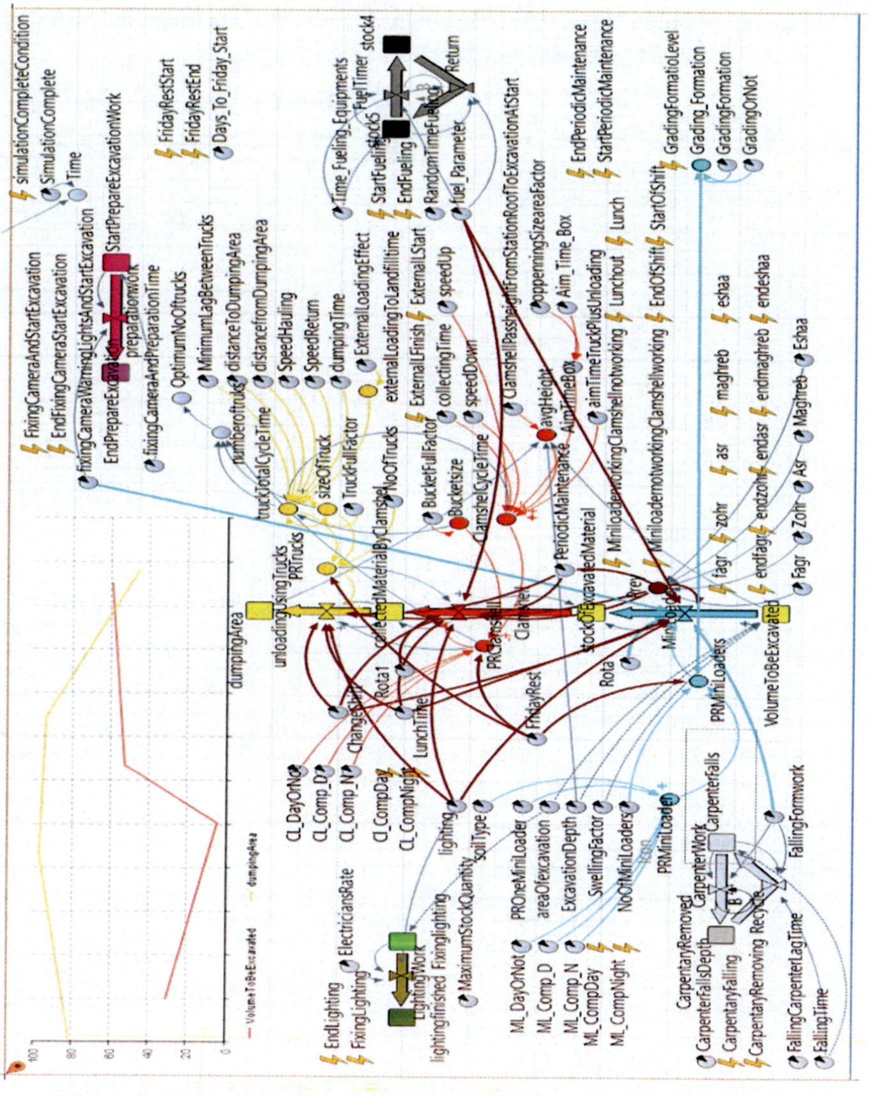

Fig. 2. Extracted from AnyLogic simulation programme to represent the model

8 Testing and Validation

The proposed model was tested and validated through results comparison of actual time to finish the operation to the model time output. Table 4 presented a comparison between the model's outcome to the actual cases via the mean value of time finishing the operation acquired by the simulation model for each excavation box at 10000 iterations.

Table 3. Example of parameters value for box A11 from Ticket to Intermediate level (Data entry of excavation box to AnyLogic experiment).

Parameter Name	unit	D OR P	Parameter value
Number of Mini Loaders	By unit	D	2
Soil type	ratio	D	1
Mini Loaders operator competent factor day shift	ratio	D	0.85
Mini Loaders operator competent factor night	ratio	D	0.9
PR of One Mini Loader	M3/sec	P	triangular(0.03,0.05,0.04)
Grading formation factor	ratio	P	triangular(0.5,0.7,0.6)
Clamshell bucket size	M3	D	2.6
Bucket speed up	Km/hour	D	10
Clamshell aim time box factor	hour	P	triangular(0.0015,0.003,0.0025)
Lag between trucks time	hour	D	0.002
No of trucks	By unit	D	4
Carpentry falls time	hour	P	max(1, normal(1,3))
Lighting work rate	M2/hour	D	150
Area of excavation	M2	D	632
Excavation depth	m	D	7.8
Bucket speed down	Km/hour	D	25
Bucket pass from roof to top of excavation at start	m	D	4+7.5
Clamshell aim time truck and unloading the	hour	P	min((max((normal(0.0002,0.003)),.002))
Clamshell Bucket collecting time	hour	P	min((max(
Clamshell competent operating factor day shift	ratio	D	0.9
Clamshell competent operating factor night shift	ratio	D	0.95
Distance from excavation location to dumping	km	D	0.8
Truck Hauling speed	Km/hour	P	triangular(8,13,10)
Truck dumping Time	hour	P	triangular(.008,.012,.01)
Preparation time before start excavation	hour	P	triangular(4.5,6.5,5)
Falling carpenter depth	m	D	3.25
Distance from Dumping area to excavation	km	D	1.3
Truck Return speed	hour	P	triangular(10,15,13)
External Loading effect time during night shift	hour	D	0.02
Fueling Time	hour	P	uniform(1,10)
Temp opening size factor	ratio	D	1
Truck Fulling Factor	hour	P	triangular(0.9,0.97,0.94)
Swelling Factor	ratio	D	1.12
Stock quantity of excavated material to be collected by Mini Loaders	M3	D	60
Bucket full factor	ratio	P	triangular(0.75,1,0.96)
Size of truck	M3	D	15
Time Fueling equipments	hour	D	0.75
Time Collecting Falling formwork	hour	P	triangular(0.25,0.6,0.5)
Grading depth	ratio	P	triangular(0.5,0.7,0.6)

Through comparing the mean data proposed from the model experiments to the field actual data it can be noticed that the production rate estimated from this model is approximately 1.06 times the filed actual production rate, and the variance between field and model equivalent production rates are vary (from −1.6% to +13.2%). Using mean square error shows that the error in these results is equal to 4.14%, which is a reasonable error value. It is noticed that mostly the model PR is greater than the field actual PR. This is due to the fact that the model is may be missing some additional factors or parameters not taken in consideration like adverse weather condition and labor and operators fatigue or congestion between mini-Loaders effect, which will be addressed in future research to tackle the current assumptions.

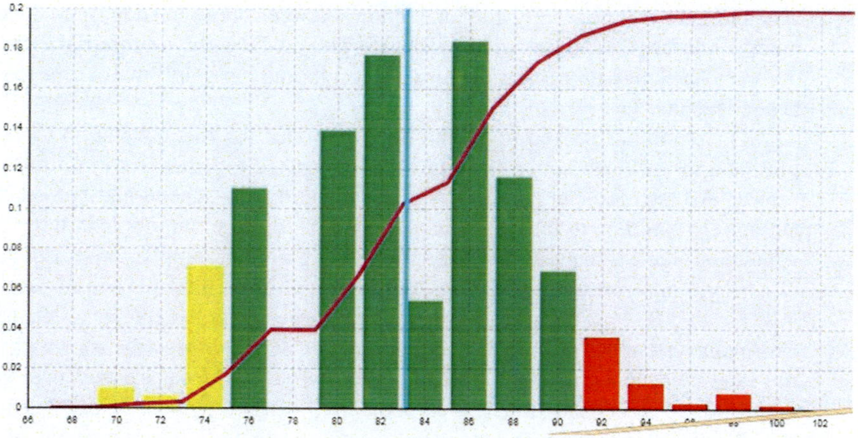

Fig. 3. Output of Monte Carlo experiment run for box A11 of 10000 runs to show the model time (in hours) required to finish the whole excavation operation, mean value of the model is (83.15 h)

Table 4. Comparison between Equivalent PRs for Model and field PRs to all excavation boxes.

Box Title	**Volume** to be excavated and hauled (m3)	Total actual time of **Field Results** (working hours)	**Mean** of Total Time finishing the job from **Model** (working time in hours)	**Equivelant PR** extracted from **Actual** Results (m3/hour)	**Equivelant PR** extracted from **Model** (m3/ hour)	**(PR Model / PR Actual)** %	(PR Model -PR actual) / PR actual %	(PR Model -PR actual)^2
11A	4,929.6	92	83.15	53.6	59.3	110.6	10.6	32.5
10A	2,386.8	48	42.42	49.7	56.3	113.2	13.2	42.8
1,2A	6,583.2	104	105.73	63.3	62.3	98.4	-1.6	1.1
8,9A	5,709.6	108	95.73	52.9	59.6	112.8	12.8	45.9
3A	3,260.4	58	55.45	56.2	58.8	104.6	4.6	6.7
13C	6,909.5	123.5	116.81	55.9	59.2	105.7	5.7	10.3
14,15,16C	7,370.4	129	123.49	57.1	59.7	104.5	4.5	6.5
1,2C	4,605.6	79.5	79.88	57.9	57.7	99.5	-0.5	0.1
3C	7,650.1	144	128.70	53.1	59.4	111.9	11.9	39.9
cc1	6,060.5	104.5	103.03	58.0	58.8	101.4	1.4	0.7
cc2	6,113.2	106	103.63	57.7	59.0	102.3	2.3	1.7
							Σ =	188.1
							Mean square error % $\sqrt{(\Sigma / n)}$ =	4.14

9 Conclusion and Future Work

The research aims at providing more accurate production rate estimation of construction activities using SD simulation modeling technique based on interactions and heterogeneity of crews. This approach helps to determine more realistic schedules and provide managers and decision makers with more accurate predictions. The research utilizes excavation using clamshell in metro subway as a case study. Although the excavation has a lot of complexity, SD has proved able to simulate and replicate the process and accounting for heterogeneity and interactions between crews in addition to the dynamic nature of process. In addition, the model has proved the ability to forecast PR with high accuracy level that quite matching the field results and it requires minor time to get these desired results.

SD method has proved to be an effective production rate estimation tool as it can present a lot of construction activities that connected together to form a big part of a project. As such SD can help in stochastic planning and scheduling. Nonetheless, the authors are planning to further develop the model to address multiple issues. One basic simulation like SD may be not enough for representing complex processes like most of construction work activities. Thus, combining all three methods of simulation (DES, ABM, and SD) into the model will provide more accurate representation of reality. The main system can still be SD but DES can be added to describe trucks transferring the excavated material to the dumping area, while ABM can be added to show the interactions between individuals at carpentry removing crew, electricians' lighting work, mini-Loaders congestion, and to show the fatigue effect of crew and operators.

Also, the presented model will be tested for more than these 11 excavation boxes at different stations with other circumstances to verify this model. Finally, the SD model can be used to determine the optimum production rate based on parameters variation and different crew constituents.

References

AbouRizk, S.: Role of simulation in construction engineering and management. J. Constr. Eng. Manage. **136**(10), 1140–1153 (2010)

AbouRizk, S., Knowles, P., Hermann, U.R.: Estimating labor production rates for industrial construction activities. J. Constr. Eng. Manage. **127**(6), 502–511 (2001)

AbouRizk, S.M., Halpin, D.W.: Statistical properties of construction duration data. J. Constr. Eng. Manage. **118**(3), 525–544 (1992)

Bokor, O., Florez, L., Osborne, A., Gledson, B.J.: Overview of construction simulation approaches to model construction processes. Organ. Technol. Manage. Constr. Int. J. **11**(1), 1853–1861 (2019)

Borshchev, A.: The Big Book of Simulation Modeling: Multimethod Modeling with AnyLogic 6. AnyLogic North America, Chicago (2013)

Borshchev, A., and Filippov, A.: From system dynamics and discrete event to practical agent based modeling: reasons, techniques, tools. In: Proceedings of the 22nd International Conference of the System Dynamics Society, vol. 23 (2004)

Chong, W.K., O'Connor, J.T., Chou, J.-S., Lee, S.-H.: Predicting the production rates of foundation construction using factor and regression analysis. In: Construction Research Congress 2005, American Society of Civil Engineers, San Diego, California, pp.1–10 (2005)

Christian, J., Hachey, D.: Effects of delay times on production rates in construction. J. Constr. Eng. Manage. **121**(1), 20–26 (1995)

Craig, W.K.: System Dynamics Methods A Quick Introduction (1998)

El-Maghraby, E., Elshakour, H.A.: Predicting the production rate of pouring ready mixed concrete using regression analysis. J. Civil Eng. Sci. **3**, 16 (2014)

Ergan, S.K., Akinci, B.: Automated approach for developing integrated model-based project histories to support estimation of activity production rates. J. Comput. Civil Eng. **26**(3), 309–318 (2012)

Gerald, H.: How to Calculate Production Rates? (2019). http://www.Bizfluent.html

Grigoryev, I.: AnyLogic 7 in Three Days: A Quick Course in Simulation Modeling. Verlag nicht ermittelbar, Erscheinungsort nicht ermittelbar (2016)

Heravi, G., Eslamdoost, E.: Applying artificial neural networks for measuring and predicting construction-labor productivity. J. Constr. Eng. Manage. **141**(10), 04015032 (2015)

Sarvaiya, M.: What is the difference between productivity and production rate? (2019). http://www.Quora.html

Sterman, J.D.: System Dynamics: Systems Thinking and Modeling for a Complex World. Massachusetts Institute of Technology Engineering Systems Division Working Paper Series, Cambridge, p. 32 (2002)

Ulubeyli, S., Kazaz, A., Er, B.: Planning Engineers' Estimates on Labor Productivity: Theory and Practice. Procedia Soc. Behav. Sci. **119**, 12–19 (2014)

Watkins, M., Mukherjee, A., Onder, N., Mattila, K.: Using agent-based modeling to study construction labor productivity as an emergent property of individual and crew interactions. J. Constr. Eng. Manage. **135**(7), 657–667 (2009)

Woldesenbet, A., Jeong, D.H., Oberlender, D.: Daily work reports-based production rate estimation for highway projects. J. Constr. Eng. Manage. **138**(4), 481–490 (2012)

Stepped Loading of a Raft Foundation Supported on Deep Soft Clay Improved with Floating Stone Columns

Khaled Hussein Elmeligi[1(✉)], Mamdouh A. Sabry[1], and Waleed El-Sekelly[2]

[1] Public Work Engineering, Faculty of Engineering, Cairo University, Giza, Egypt
khaledelmeligi@gmail.com

[2] Structural Engineering Department, Mansoura University, Mansoura, Egypt

Abstract. Soil improvement techniques are necessary in construction on weak soils such as soft clays and loose sands, as these weak deposits usually suffer from excessive settlement, and low bearing capacity. Stone columns are one of the most effective and economic systems to improve soils by alteration of soil in-situ properties through the installation of columns composed of granular material. Generally, stone columns are constructed to penetrate the soft soil layer and rest on deeper strong soils (known as end bearing stone columns). However, for deep soft clay layers, end bearing stone columns are not cost effective, in which case using floating stone column is a better solution. This paper includes a benchmark problem of loading a raft foundation of a tank rested on a deep soft clay layer improved using a group of floating stone columns. The 3D numerical program PLAXIS 3D was adopted to investigate the effect of raft loading technique on the final consolidation settlement value and time. Two loading techniques were used, the first is by filling the tank using staged loading technique (slow loading), and the second by filling the tank using an immediate loading technique (quick loading). This benchmark example provide an understanding of the performance of floating stone columns when loaded in a staged loading as compared to quick loading.

1 Introduction

Deep soft and very soft clay deposits usually produce excessive consolidation settlement values, low bearing capacity and long time for consolidation to occur. To avoid these problems, it is necessary to provide an improvement technique for soft soils before or during the construction stages. Many improvement techniques can be used, such as dynamic compaction, wick drains, sand drains, jet grouting, deep densification and soil reinforcement. Stone column technique is considered one of these techniques and has been used as an effective technique to improve the soft soil.

This technique depends on replacing parts of the soft soil with granular material columns (piles) by means of vibration or other construction methods to make a stiffer system with the surrounding soft soil. This will help to prevent the previously mentioned problems of soft soils to occur, as it increases the bearing capacity by improving the shear

H. Shehata and M. Badr (Eds.): *Advancements in Geotechnical Engineering*, SUCI, pp. 32–43, 2021.
https://doi.org/10.1007/978-3-030-62908-3_3

strength of soft soil, it also reduces the consolidation settlement and provide acceleration for the required time of final consolidation of the system. So; not only it has the ability to improve soft soil, but also it is considered as safe and economic technique.

Generally, stone columns are constructed to penetrate the soft soil layer to rest on stiffer layer below the soft one, this technique is known as end bearing columns. If we have a site with deep soft soil, using end bearing columns will not be economic and make it susceptible to bulging failure. So, floating stone columns technique is preferred to be used in this case. The composed system of floating column surround by soft soil is not well understood (Gab et al., 2007), there is a lack of the theoretical background in design of floating stone columns.

Although many empirical and analytical methods for stone column composite system have been conducted over the past years (e.g. Goughnour and Bayuk, 1979; Balaam and Booker 1981; Priebe 1995; Xie et al. 2009), there are no current design methods that incorporate the effect of loading technique for deep soft soil layer improved with floating stone columns. More studies need to be adopted to accurately predict the effect of floating stone columns rested on deep soft soils on the consolidation settlement, rate and consolidation time.

2 Numerical Program

This paper includes a benchmark problem of loading a raft foundation of a tank rested on a deep soft clay layer improved using a group of floating stone columns. The 3D numerical program PLAXIS 3D was adopted to investigate the effect of raft loading technique on the final consolidation settlement value and time. Two loading techniques were used, the first is by filling the tank using staged loading technique (slow loading), and the second by filling the tank using an immediate loading technique (quick loading).

2.1 Model Geometry

Different sizes of tank raft with breadth (B) and thickness (t) rests on a deep soft clay layer with thickness (H). Below this soft clay layer is a dense sand layer with thickness (Hs). A 0.5 m sand cushion was used as a blanket just below the foundation to enable the excess pore water pressure do dissipate during consolidation, and placed over the site area which beyond the model boundary.

The deep soft clay layer was improved by floating group of stone columns beneath the sand cushion, the stone columns diameter (D) was chosen to be equal 1.0 m as this is a practical value used in various sites, the stone columns are with length (L) with a square pattern of spacing (S). An additional row of stone columns in the X and Y directions was added just after the edge of the raft to prevent deformation and avoid any influence of the outer boundary of the raft. Only one-quarter of the models was simulated due to symmetry and to avoid time consuming. Figure 1 indicates the geometric numerical model. Table 1 presents the geometric studied parameters.

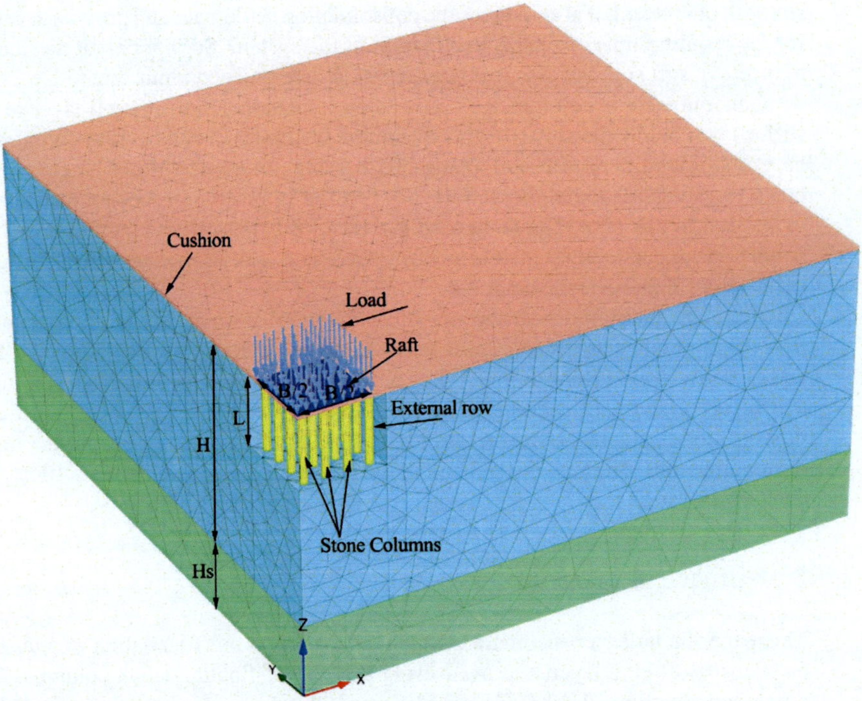

Fig. 1. Geometric modelling of the improved case

Table 1. Geometric properties

Raft Width (B) m	Raft Thickness (t) m	Soft Clay (H) m	Dense Sand (Hs) m	S.C Length (L) m	S.C Diameter (D) m	S.C Spacing (S) m
20.00	0.75	30.00	10.00	10.00	1.00	2D 3D
30.00				15.00		
40.00				20.00		

2.2 Loading Technique

Generally, in practice when using stone columns to improve soft soil layers, a pre-loading phase with an ultimate load is performed. Then after that, the proposed building is to be constructed. In this study, a load of 50.00 kPa represent the case where the tank is full. Two types of loading techniques were used as follows:

First Technique: (Staged Loading)
The tank was filled using an incremental technique consist of four stages each stage represent loading and consolidation phases. The fifth stage is a consolidation phase until the full dissipation of pore water pressure.

Second Technique: (Immediate or quick Loading)
The tank was filled using a quick technique consisting of one stage represent loading or filling the tank. The second stage is a consolidation phase until the full dissipation of pore water pressure.

Tables 2 and 3 present the two loading techniques used.

Table 2. First loading technique (Staged loading)

Stage	Technique	Time (Day)	
		From initial	Incremental
Stage I - (10 kPa)	Loading	0–7	Day (0–30)
	Consolidation	0–30	
Stage II - (20 kPa)	Loading	0–37	Day (30–60)
	Consolidation	0–60	
Stage III - (30 kPa)	Loading	0–67	Day (60–90)
	Consolidation	0–90	
Stage IV - (50 kPa)	Loading	0–100	Day (90–100)
Stage V - Final Consolidation	Consolidation	∞	Day (100 – End of Consolidation)

Table 3. Second loading technique (Quick loading)

Stage	Technique	Time (Day)	
Stage I – (50 kPa)	Loading	0–7	Day (0–7)
Stage II – Final Consolidation	Consolidation	∞	Day (7– End of Consolidation)

2.3 Soil Material Properties

Soft Clay Layer
A special feature of soft soils is their high degree of compressibility, the best soil model for these types of soils is Soft Soil Model. This soil model has the following features: Stress dependent stiffness, can memorize pre-consolidation stress, failure behavior according to the Mohr-Coulomb criterion, distinguishing between primary loading and unloading-reloading, realistic representation of soil composite behavior, and representation of creep and secondary compression.

Parameters of the soft clay layer used in this research were indicated in Site Investigation Report Turbine No. 3 by A&A Consultant (2015) which present the nature of the soft clay layer founded in Kafr El-Sheikh governorate, north of Egypt. For change of permeability value C_k, it was taken as $C_k = 0.50 \times e_o$ (Tavenas, F. et al. 1983). An interface value R_{int} is used for the soft clay layer to present the real slippage and gapping or overlapping effect between stone columns and the surrounding soil, it may be assumed to equal to 2/3 and don't exceed a value of 1.00.

Dense Sand Layer
The hardening soil model in PLAXIS 3D was chosen to use for the dense sand layer. The dense sand layer parameters used in this research were taken from Hafez (2003) axisymmetric FEA model. This layer was to represent a stratum below the deep soft clay layer slightly affected by the consolidation and to let the drainage path to be double drainage path.

Stone Column
The material model of the stone column material is the same as the dense sand layer which is the hardening soil model. Also, the parameters of the gravel used in this research were indicated in Hafez (2003) axisymmetric FEA model.

Tank Raft
The raft presented in this study was modelled as a plate element. Although using volume element is more realistic than using a plate element, we used a plate element to ease the process of running the model and to take less time.

Tables 4 and 5 present the soil material parameters used in the model.

Table 4. Material properties of tank raft

Parameters	Name	Raft	Unit
Thickness	d	0.75	m
Unit weight	γ	25	kN/m^3
Type of behavior	Type	Linear, Isotropic	–
Young's modulus	E_1	2×10^7	kN/m^2
Poisson ratio	υ_{12}	0.15	–

Table 5. Soil material properties

Parameter	Name	Soft clay	Dense sand	Stone column	Unit
General					
Material model	Model	Soft Soil	Hardening soil	Hardening soil	–
Drainage type	Type	Undrained (A)	Drained	Drained	–
Soil unit weight above phreatic level	γ_{unsat}	15	17	18	kN/m^3
Soil unit weight below phreatic level	γ_{sat}	17	20	21	kN/m^3
Initial void ratio	e_{init}	1.20	0.50	0.50	–
Parameters					
Secant stiffness in standard drained triaxial test	E_{50}^{ref}	–	6×10^4	9×10^4	kN/m^2
Tangent stiffness for primary oedometer loading	E_{Oed}^{ref}	–	6×10^4	9×10^4	kN/m^2
Unloading/ reloading stiffness	E_{ur}^{ref}	–	300×10^3	450×10^3	kN/m^2
Power for stress-level dependency of stiffness	m	–	0.5	0.50	–
Failure ratio	R_f	–	0.90	0.9	–
Poisson ratio	υ'_{ur}	0.30	0.20	0.20	–
Compression index	C_c	0.30	–	–	–
Swelling index	C_s	0.03	–	–	–
Cohesion	C'_{ref}	1.0	0	0	kN/m^2
Friction angle	ϕ'	26	43	44	°
Dilatancy angle	ψ	0	0	0	°
Flow parameters					
Horizontal permeability (x-direction)	k_x	1.728×10^{-3}	1	2	m/day
Horizontal permeability (y-direction)	k_y	1.728×10^{-3}	1	2	m/day

(*continued*)

Table 5. (*continued*)

Parameter	Name	Soft clay	Dense sand	Stone column	Unit
Vertical permeability	k_z	0.864×10^{-3}	1	1	m/day
Change in permeability	c_k	0.60	1×10^{15}	1×10^{15}	–
Interfaces					
Strength reduction factor	R_{inter}	0.70	1	1	–
Initial condition					
Over-consolidation ratio	*OCR*	1.50	1.00	1.00	–

3 Numerical Analysis

To study the effect of loading techniques, the square pattern grid of floating stone column with lengths L = 10, 15, 20 m and spacing S = 2D & S = 3D are adopted in the model. Different sizes of tank raft ranges from B = 20 m to B = 40 m were used, the thickness of the raft is 0.75 m. The soft clay layer thickness H = 30.0 m, the dense sand layer thickness Hs = 10.0 m, the sand cushion at the top of model is 0.50 m thickness. The two loading techniques indicated in Tables 2 and 3 were used to show the effect of loading type on the improved system.

As an example to illustrate the results, the deformed mesh resulting from applying a stress level equal to 50 kPa on the tank raft loaded with the first loading technique and resting on soft clay layer improved with floating group of stone columns for the case of B = 20 m, S = 3D and L = 15 m is presented in Fig. 2.

Curves relating the consolidation settlement versus the consolidation time for the two types of loading techniques are shown in Figs. 3, 4, 5, 6, 7 and 8. Each chart is for a specific stone column length and spacing.

For all cases, the final consolidation calculation phase stops when the maximum absolute excess pore pressure is below the prescribed value of $|P - Stop| = 1.00$ kN/m^2, this means that the predicted required time for final consolidation is calculated when the maximum excess pore pressure becomes less than the value of 1.00 kN/m^2. In other words, approximately the full dissipation of pore water had occurred.

From Figs. 3, 4, 5, 6, 7 and 8, it is noted that the final consolidation settlement value is almost the same value for the two types of loading at each foundation width (B), the main difference between the two loading techniques implies the rate of consolidation settlement. To illustrate this main difference, a curve for foundation width B = 20 m, stone column length L = 10 m, and spacing S = 2D are adopted. The X-axis in this curve express the square root of consolidation time, the Y-axis is the same as previous charts which is the consolidation settlement.

For the first loading technique, four lines are drawn for each loading phase to link each loading phase with the final consolidation one, these lines are labelled as L_1 to

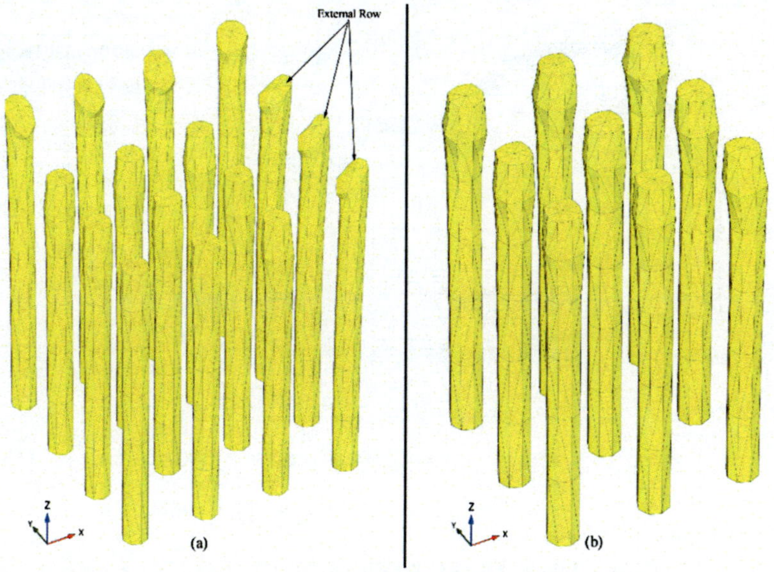

Fig. 2. Deformed mesh for stone columns (a) With external row, (b) Without external row

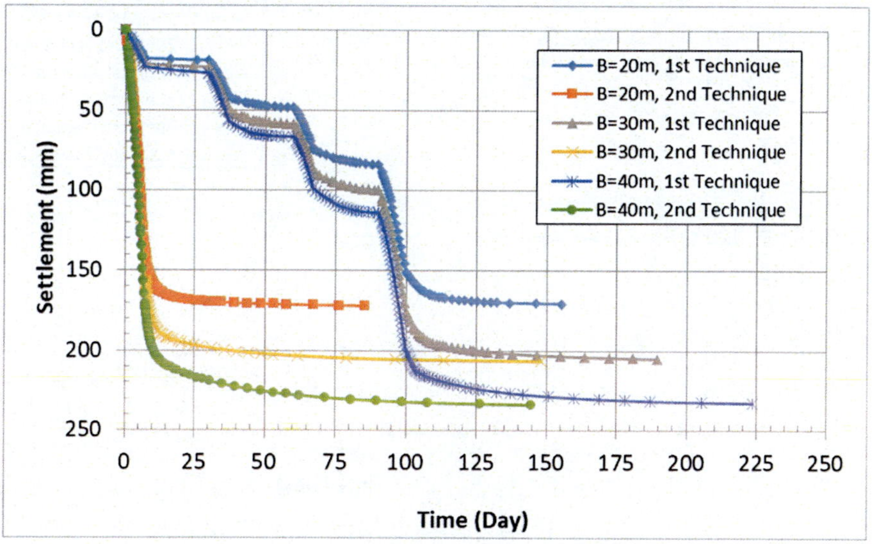

Fig. 3. Effect of Loading Technique, L = 10 m, S = 2D

L_4 respectively. For the second loading technique, one line is drawn to link the loading phase to the final consolidation one, this line is labelled as L_5.

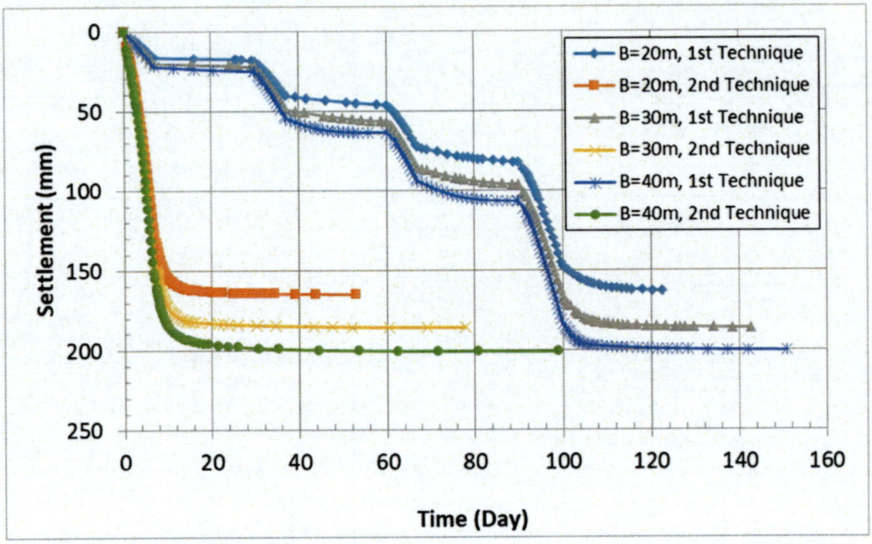

Fig. 4. Effect of Loading Technique, L = 15 m, S = 2D

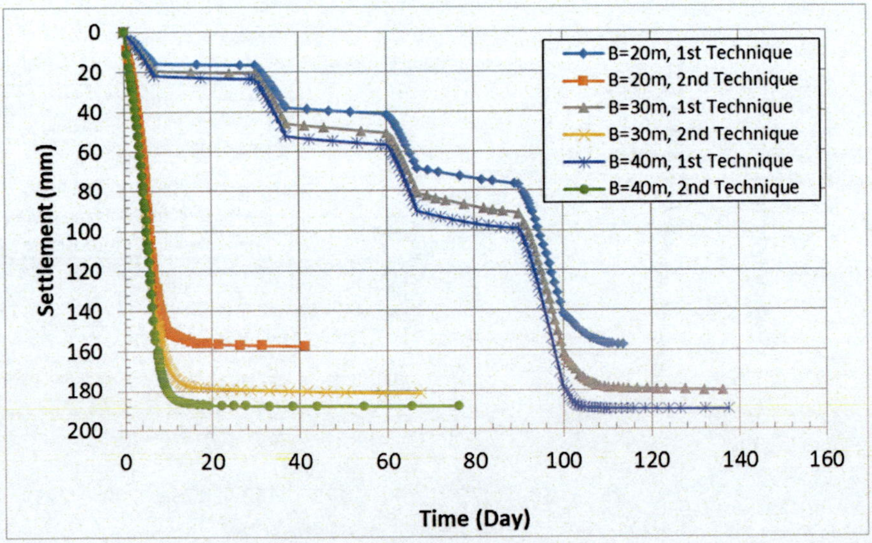

Fig. 5. Effect of Loading Technique, L = 20 m, S = 2D

The rate of settlement in each loading phase for both two types of loading techniques is expressed as the ratio between the difference in settlement to the difference in time, which is known as the slope of each line = $\frac{\Delta S}{\Delta T}$ in mm/Day.

From Fig. 9, it is noted that slopes of each loading phase in the first loading technique (slope of L_1 to L_4) are less than the slope for loading phase in the second loading

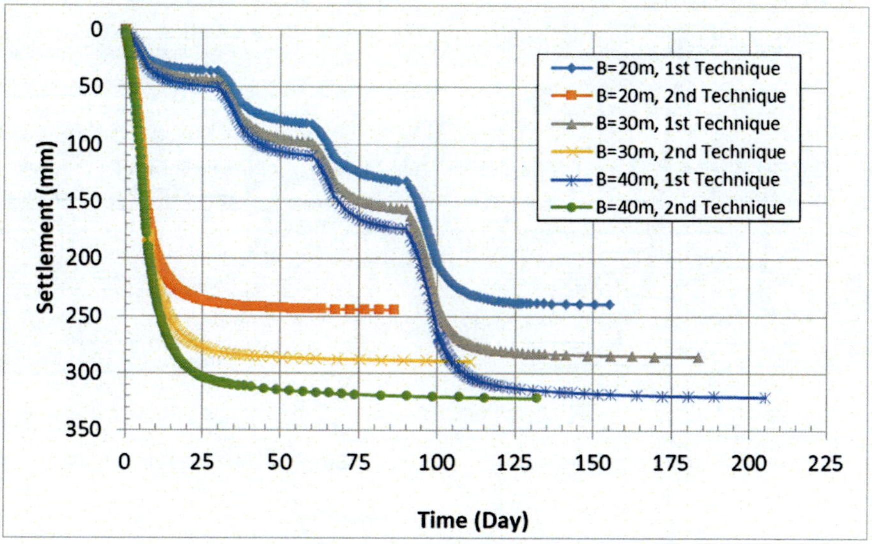

Fig. 6. Effect of Loading Technique, L = 10 m, S = 3D

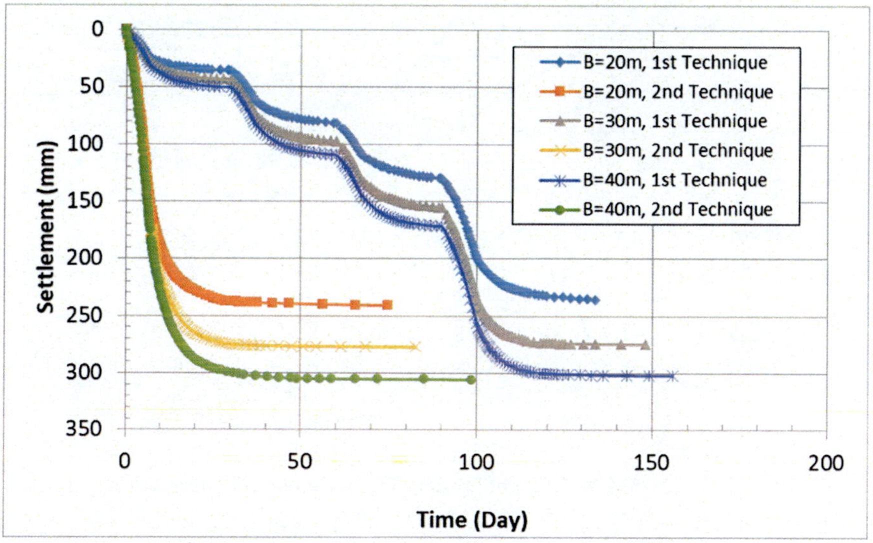

Fig. 7. Effect of Loading Technique, L = 15 m, S = 3D

technique (slope of L_5). In other words, the rate of settlement for the first loading technique (incremental) is less than the rate of settlement for the second loading technique (immediate).

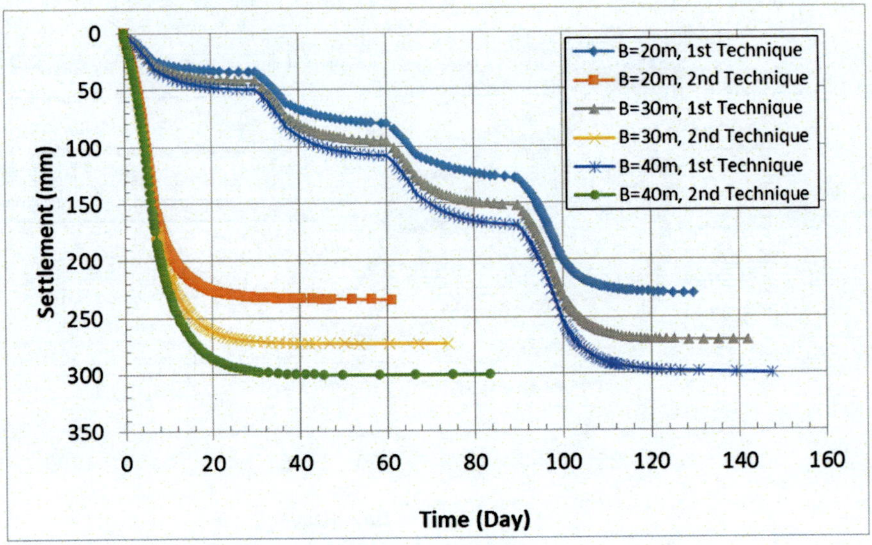

Fig. 8. Effect of Loading Technique, L = 20 m, S = 3D

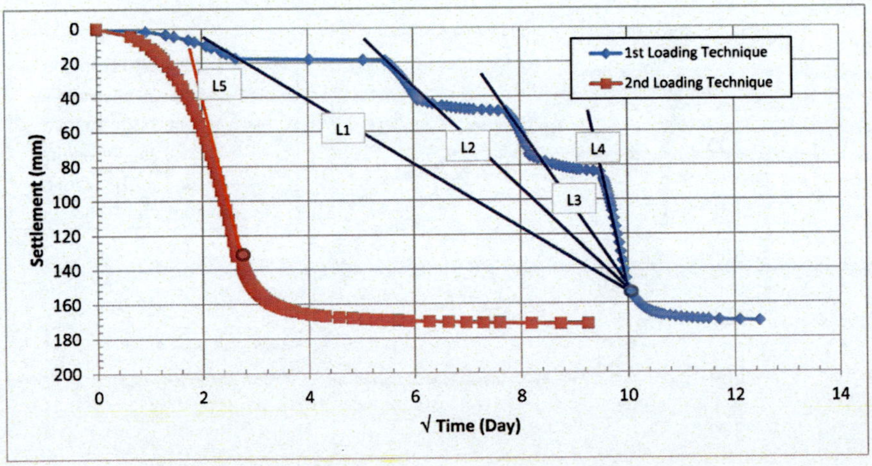

Fig. 9. Effect of Loading Technique on Rate of Settlement

The reason for this is that applying the first loading technique increases the soft clay layer stiffness, as the full load is applied in stages, and each stage allow the dissipation of excess pore pressure to occur resulting more consolidated soft clay layer, while in the second loading technique; the full load is applied once. However, both the two types of loading techniques give the same final consolidation settlement.

4 Conclusions

Using the developed numerical model as illustrated to investigate the effect of tank raft loading technique rested on deep soft clay layer improved by floating stone column group was adopted. The results of this study for each raft width and the corresponding stone columns length produce the following conclusions:

1- Changing loading technique whether incremental loading or immediate loading, doesn't affect the final consolidation settlement value. The final consolidation settlement value is approximately equal for both of the two loading techniques.
2- The rate of settlement for the first loading technique "incremental loading" is less than the rate of settlement for the second one "immediate loading".

References

Balaam, N.P., Booker, J.R.: Analysis of rigid rafts supported by granular piles. Int. J. Numerical Anal. Methods Geomechanics **5**(4), 379–403 (1981)

Gäb, M., Schweiger, H., Kamrat-Pietraszewska, D., Karstunen, M.: Numerical analysis of a floating stone column foundation using different constitutive models. Focus on Ground Improvement., pp. 137–142. Taylor & Francis, Geotechnics of Soft Soils (2008)

Goughnour, R.R., Bayuk, A.A.: A field study of long term settlements of loads supported by stone columns in soft ground. Proc. Intl. Conf. on Soil Reinforcement: Reinforced Earth and Other Techniques, vol. 1, pp. 279–286 (1979b)

Hafez, D.: Finite Element Study of the Behavior of Stone Columns, Full Thesis (2003)

Priebe, H.J.: The design of vibro replacement. Ground Eng. **28**, 31–37 (1995)

Tavenas, F., Jean, P., Leblond, P., Leroueil, S.: The permeability of natural soft clays, part ii: permeability characteristics. Canadian Geotechn. J. **20**, 645–660 (1983)

Xie, K.H., Lu, M.M., Liu, G.B.: Equal strain consolidation for stone columns reinforced foundation. Int. J. Numer. Anal. Methods Geomech. **33**(15), 1721–1735 (2009a)

Xie, K.H., Lu, M.M., Hu, A.F., Chen, G.H.: A general theoretical solution for the consolidation of a composite foundation. Comput. Geotech. **36**(1–2), 24–30 (2009b)

The Use of a Robotized Inclinometer System to Measure Deep-Seated Ground Deformation in a Monumental Area During TBM Tunnel Excavations. The Case of Rome Subway, New Line C

Paolo Allasia[1], Danilo Godone[1](✉), Giorgio Pezzetti[2], Ivan Mammone[3], and Eliano Romani[3]

[1] Research Institute for Geo-Hydrological Protection (CNR-IRPI) Turin, National Research Council of Italy, 10135 Turin, Italy
{paolo.allasia,danilo.godone}@irpi.cnr.it
[2] SMAK S.a.s., Bergamo, Italy
gpezzetti@smaksas.eu
[3] Metro C S.c.p.A., Rome, Italy
{i.mammone,e.romani}@metrocspa.com

Abstract. The new section T3 of Rome Metro network (line C) stretches in an important World Heritage area (Colosseo, Fori imperiali etc.). In order to monitor and assure the safety of this heritage, a very complex monitoring network has been installed to provide information on monuments status, on the effectiveness of the designer's hypothesis and on the evolution of the TBM tunnel excavation. In cooperation with METROC S.c.p.A, General Contractor for planning and construction of the project, a robotized inclinometer system was experimentally installed with the aim of measuring the deep-seated ground deformations in three sections with different inclinometer tubes length. This equipment allows the measurement of deep horizontal displacements using a single inclinometer probe and a robotic system that automatically manage the probe for the entire length of the borehole with 50 cm step and double reading (0/180°). The experimental activity was carried out before, during and after the TBM excavation and allowed the monitoring of three sections at the Basilica of Santo Stefano Rotondo al Celio (62 m), Fori Imperiali (43 m and 33 m) with a frequency of measurement up to 6 measurements/day. The results showed interesting relations between TBM progress and horizontal deep deformations induced by tunnel excavation. These deformations, although very small, highlighted the potential of this instrumentation in terms of accuracy, vertical spatial resolution and very high frequency of measurements. This type of results can be useful to a more complete interpretation of the subsoil deformations and interaction between excavation and the surrounding built environment.

H. Shehata and M. Badr (Eds.): *Advancements in Geotechnical Engineering*, SUCI, pp. 44–58, 2021.
https://doi.org/10.1007/978-3-030-62908-3_4

1 Introduction

Section T3 of Line C (Rome Metro) passes through the historic center in a context characterized by numerous historical monuments and World Heritage Sites (e.g. Colosseo, Basilica di Massenzio, Basilica di Santo Stefano Rotondo al Celio, Santa Maria in Domnica, Fori Imperiali). The complexity of the area required the establishment of a complex automatic and manual monitoring network, capable of providing designers and work manager, dataset for: i) safety and protection of the monuments, ii) validation of the design hypotheses, iii) assessment of the impact of the works (Fig. 1). The monitoring activities concerned the soil and subsoil as well as the monuments and items located at a certain distance of potential influence of the works. Where deemed significant, temporary support structures were specifically installed in order to mitigate the deformation or failure. Also these temporary structures were monitored. The investigated time span covered a period of at least one year before the operation and will continue for at least one year from the completion of the works.

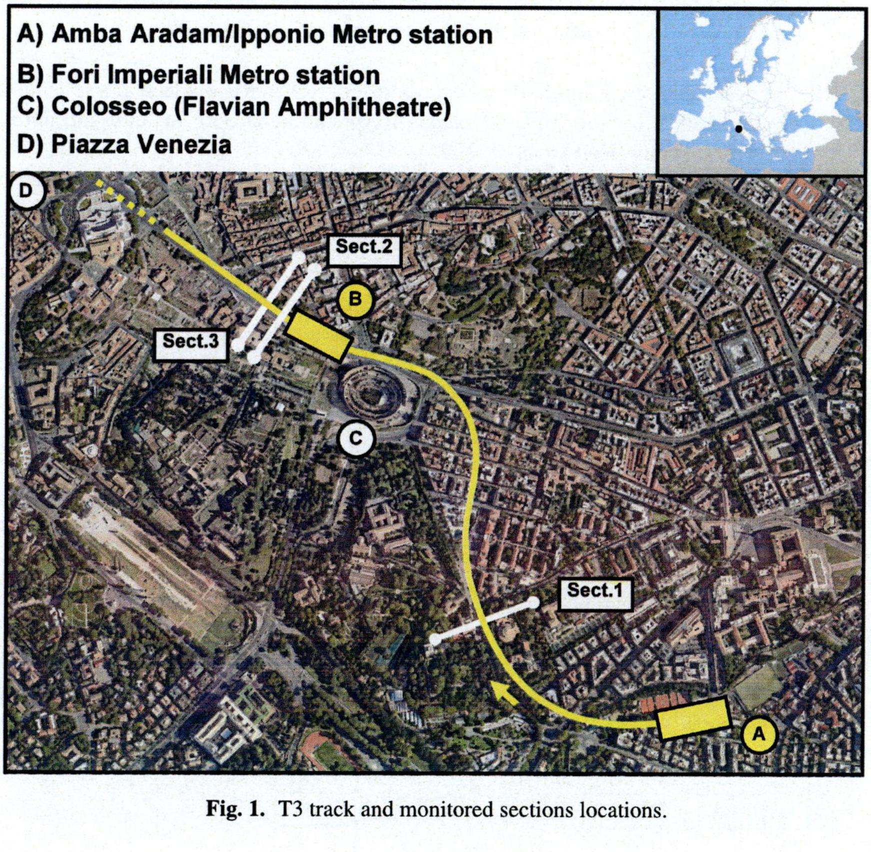

Fig. 1. T3 track and monitored sections locations.

In this context, METRO C and CNR-IRPI have collaborated in the experimentation of an innovative system for measuring horizontal deep-seated ground displacements for the evaluation of the deformations induced by the mechanized excavation activities with TBM. The system, developed and patented by the CNR IRPI, has been widely used for monitoring landslides but had never been used in the context of large scale infrastructure projects (Lollino et al. 2002, 2006).

This system, called Automated Inclinometer System (AIS) was set up with the aim of performing the traditional inclinometer measurements in a robotic way. The peculiarities of the system are:

- use only one inclinometer probe;
- very high accuracy and repeatability;
- double reading approach (0/180°);
- high vertical spatial resolution (0.5 m);
- same equipment for inclinometer tubes with length up to 120 m;
- high frequency of measurement (up to 6 ÷ 8 measurements per day);
- possibility of rapid transfer and reuse of the instrument (following the excavation progress or in case of too large tube deformations in landslides)

These features, combined with remote control, have made it possible to use the instrumentation for the study of instability phenomena both in Italy and abroad with interesting results (Herrera et al. 2017).

As part of the monitoring of the T3 section, interest in instrumentation arose in anticipation of the excavation of the Amba Aradam/Ipponio - Fori Imperiali sub-section and in particular for the deeper section (about 55 m) near the Basilica of Santo Stefano Rotondo al Celio and Santa Maria in Domnica (Fig. 1, 2). The objective of monitoring was an accurate assessment of the potential deformations at the various depths also in consideration of the rapid advancement of the excavation front (up to 8 ÷ 10 m/day). This sector was in fact the beginning of the excavation of one of the most delicate archaeological areas of the city, and the World, and therefore all the best technologies available in terms of excavation and monitoring techniques were adopted. Following the encouraging results obtained in this section, the instrumentation was progressively transferred and

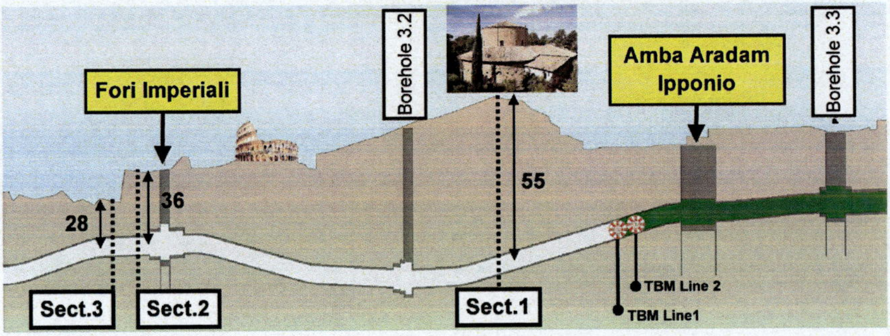

Fig. 2. Profile sketch of T3 track.

used in two further sectors located immediately after of the Fori Imperiali station near the odd (Sect. 2) and even (Sect. 3) tracks respectively (Fig. 3). Overall, approximately 750 measurement cycles were performed with double reading (equal to approximately 35 km of measured tube) with an acquisition frequency that was modulated according to the progress of the tunnel excavation (1 ÷ 6 per day)

Fig. 3. Installation of the same instrument, in a fiberglass case, in the three sites: Basilica di Santo Stefano Rotondo al Celio (Section 1); Via dei Fori Imperiali (Section 2); Via dei Fori Imperiali/Basilica di Massenzio (Section 3). In the subfigure 1, 2 the instrument installation and setup.

2 Methods and Technologies

AIS can be defined as a robotic system capable of performing inclinometric measurements following the same measurement principles adopted for manual measurements by an operator. Measurements are performed during the ascent phase with 0.5 m steps and following the known double reading approach (0°, automatic probe rotation, 180°). The descent (continuous) and the ascent (per step) are managed by a robotic system called Ground Control Unit (GrCU) which combines the use of a micromotor and a high precision rotary encoder (Allasia et al. 2018).

From the point of view of measurement sensors, the system uses a standard biaxial inclinometric probe, equipped with servo-accelerometric or MEMS technology sensors (Stark et al. 2008). For the described study, a probe with MEMS sensors from the OTR company (www.otr-geo.com) was used. In traditional manual systems, the probe is connected to the datalogger placed on the surface, by means of a special graduated cable. For the AIS, an Inclinometer Control Unit (InclCU) was developed thus allowing to replace the traditional cable with a thin synthetic fiber cable (Dyneema®, ϕ2 mm) for the only mechanical support of the probe during the measurement phases (descent/ascent) or standby. The InclCU (Fig. 4b) directly connected with the inclinometric probe, is able to powering the probe and analyzing its inclinometric signals and therefore determine whether it is moving (during the descent or ascent) or stopped at a measurement point. The amplitude of the inclinometric signal in these two states (probe stopped/probe in motion) is significantly different and the InclCU is able to recognize them and store or discard the progressively acquired data (Allasia et al. 2018). This implementation has therefore made it possible to make the probe completely autonomous and independent of the movement system and to ensure that communication between them takes place only during the start or at the end of the measurement session. On such occasions, the configuration parameters (start-up phase) or the measurements stored during the step-up

Fig. 4. Main components: a) inclinometric probe; b) Inclinometer Control Unit (InclCU); c) batteries and induction charging system; d) Dyneema® cable; e) Electronic cable wrapper; f) Up/down motor; g) Rotary encoder; h) Load cell; i) Induction charger; j) 150 m Dyneema cable reel.

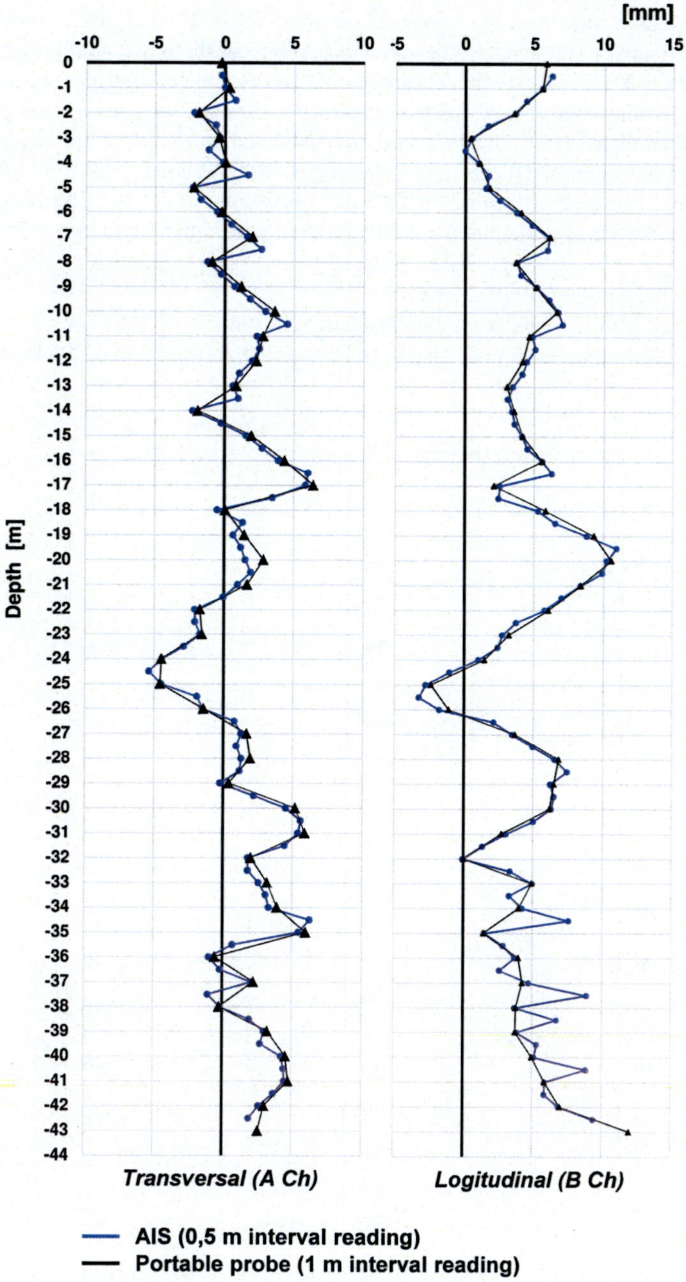

Fig. 5. Comparison of local displacement acquired in Section 2 between manual (1 m step) and robotized (0.5 m step) measurements.

phase are transferred via wireless communication. In every other measurement phase there is no communication between the probe and the movement system (Fig. 3, 4).

In order to validate the measurements and assure continuity to the monitoring activities, a comparative analysis was carried out for each section with the previous manual measurements (Fig. 5). The comparison showed an excellent correspondence between the two systems and the average differences settled at about −0.4 mm ± 0.3 for the longitudinal axis and −0.1 mm ± 0.3 for the transverse one. These differences are negligible in the engineering field and are mainly attributable to:

a) different metrological chain;
b) not identical positioning on the borehole;
c) different measurement step (1 m for manual measurements, 0.5 for automatic ones).

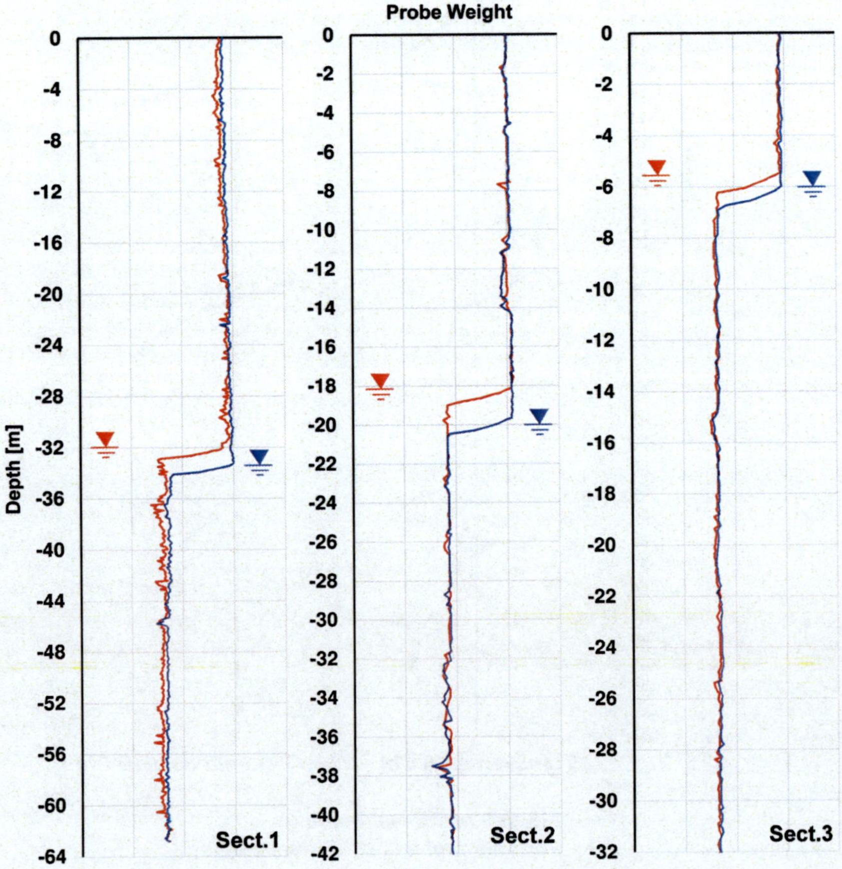

Fig. 6. Water level measured into inclinometer tube using the load cell data (in red starting day, in blue ending day of analyzed period)

As for point b), in the AIS it is possible to change the starting point but considering the already excellent correspondence of the measurements and it was not inserted. Concerning point c), the deepest part of the longitudinal axis is of particular interest, where a more marked difference between the two curves is observed. This behavior is observed only at the i-th 50 cm step measured by the robotic system and not by the one with the operator (100 cm step). During all the measurement phases, the measurement subsystem is constantly weighed by a load cell. This system aims to identify any problems that may prevent correct uplift/downlift of the probe (large tube deformations, high curvature etc.). In these cases, the system enters in a safety mode and all the measurements activities are interrupted without "losing" the probe. Recently, the load cell data has also been used to automatically identify the water level inside the tube by detecting the start of the hydrostatic thrust, which causes a noticeable reduction in weight values (Fig. 6).

3 Results

3.1 Section 1 – Santo Stefano Rotondo al Celio

The first monitoring section, located at the Basilica di Santo Stefano Rotondo al Celio, represented the beginning of the measurement activity with the robotic system. The objective, similar to that of the subsequent areas, was to acquire a robust dataset in undisturbed conditions and then continue during and after the TBM transit (Tables 1 and 2).

Table 1. Monitored sectors

Sector.	Tube length (m)	Axis depth (m)	Distance tube/excavation	Main stratigraphic layer (at excavation depth)
1	62.5	55	5	Clay
2	42.5	36	2.5	Sand
3	33.5	28	1.5	Sand

Table 2. Main parameters – Section 1

Tube length (m)	Measuring start	Measuring end	TBM transit	N° measures
62.5	06/03/2019	12/06/2019	11/05/2019	290

In general, the detected deformations (Fig. 7) fall within the instrumentation and method tolerances observed in the undisturbed period. However immediately above the outer shell, moderate (sub-millimeter) attitude changes were detected for the longitudinal axis only and were not considered particularly significant.

3.2 Section 2 – via dei Fori Imperiali

Following the interesting results obtained in the first section, the equipment was moved to section number 2 located in the Via dei Fori Imperiali construction site. The technical time for uninstalling and returning to service on the new inclinometer tube was approximately 5 h thus confirming the possibility of rapid instrument moving.

Table 3. Main parameters – Section 2

Tube length (m)	Measuring start	Measuring end	TBM transit	N° measures
42.5	25/06/2019	18/09/2019	12/09/2019	272

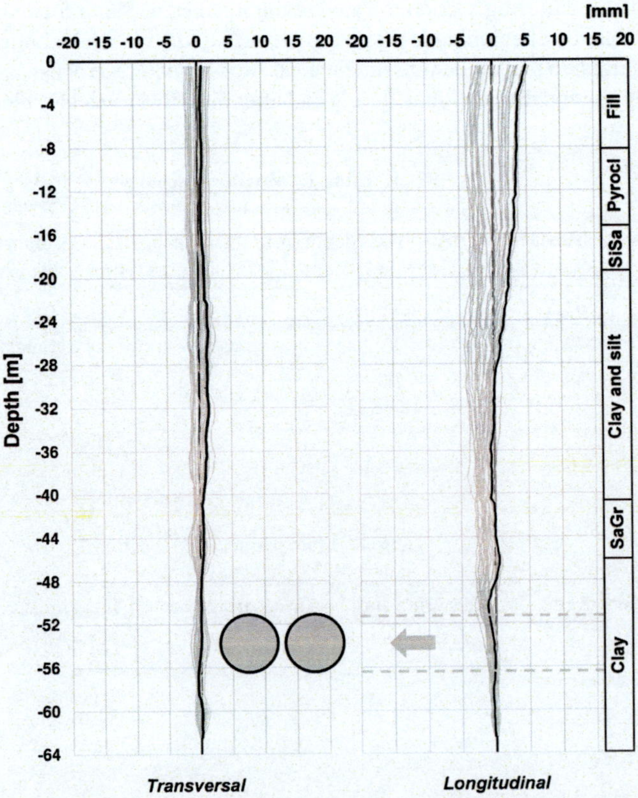

Fig. 7. Plot of cumulated displacement. For clarity sake 30 measures out of 290 are plotted (Table 2). The last measurement is highlighted in bold.

Despite the diversity of the terrains, similarly to what emerged in the previous measurement section, no substantial deformations were observed (Fig. 8). All the measurements in the orthogonal plane and in the one parallel to the excavation axis fell within the usual tolerances adopted for this type of measurement (Dunnicliff 1993; Mikkelsen 2003).

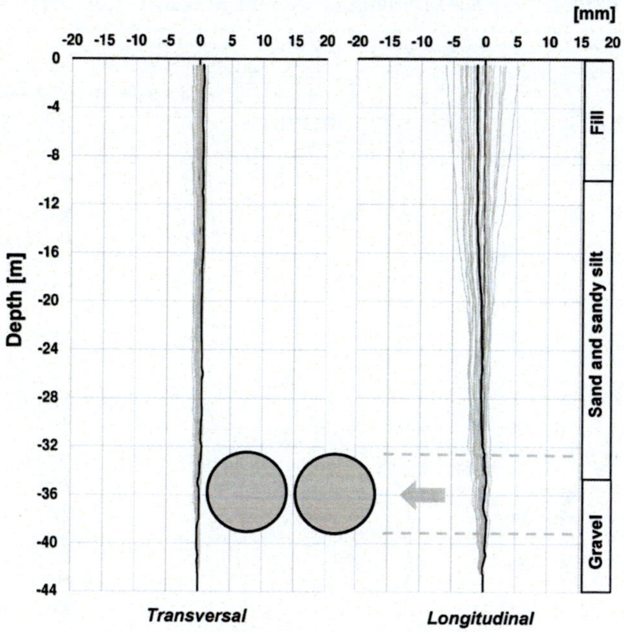

Fig. 8. Plot of cumulated displacement. For clarity sake 30 measures out of 272 are plotted (Table 3). The last measurement is highlighted in bold.

3.3 Section 3 – Basilica di Massenzio

The last analyzed sector is located very close of the Basilica di Massenzio (Via dei Fori Imperiali). This section was of particular interest as the inclinometric tube/excavation distance was the lowest of the three sections monitored (approximately 1.5 m). The soil profile are mainly sands (similar to Section 2) with a water table level located approximately 12.5 m from the ground level.

The results obtained in Sect. 3 differ considerably from the previously ones (Fig. 9). In fact, although the measured deformations are modest, around excavation depths, deformations have been observed with a clear curvilinear trend induced by the mechanized excavation in overpressure (Beghoul Demagh 2019). The development of this curvature has been observed with an increasing modality as the TBM approaches, starting from a distance of about 20 ÷ 22 m. The maximum value of the displacement, equal to about 0.5 mm, was measured roughly at the axis of the tunnel and has not undergone further changes as the TBM and processes progressed (Fig. 10). As regards the vertical extension of the perturbation, the curvilinear trend tends to run out a few meters above the top of the tunnel shell. At a depth of about 16 m, 8 m above the excavation, some minor displacements are also observed probably due to flaws in the cementation of the inclinometer tube. Concerning the longitudinal plane, no appreciable changes were observed. Similar results have been observed on the surface, where, the integration from below shows millimetric values for both measurement planes (transversal and longitudinal).

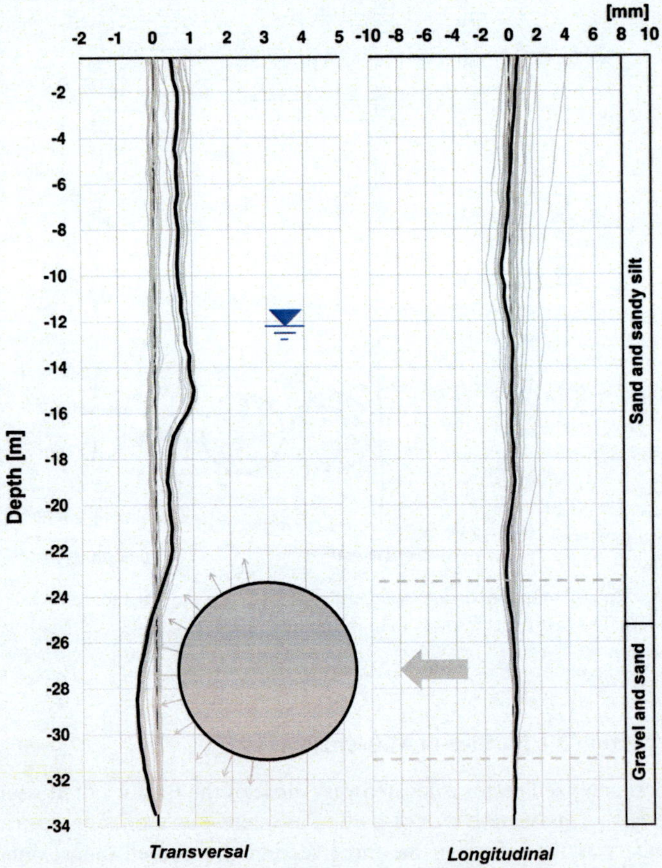

Fig. 9. Plot of cumulated displacement. For clarity sake 30 measures out of 160 are plotted from the period 5/10–26/11/2019. The last measurement is highlighted in bold.

In order to validate the noteworthy result, the inclinometric data were also compared with the acquisitions of a TRIVEC inclination-setter located at about 2 m distance. Even the data acquired with this system showed the same type of curvilinear trend with a slight overestimation in terms of modulus. Certain differences appear physiological since although very close they are not the same identical tube as the instrumentation

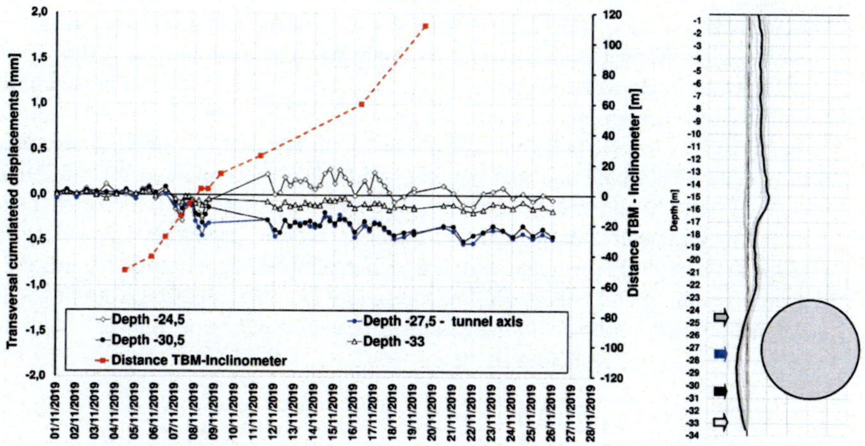

Fig. 10. Time series of transverse cumulated displacement at significant depths in terms of TBM excavation and design.

used is different. The sign, and therefore the concavity of the curve, is the same as that measured by the AIS. The measured deformations (AIS, TRIVEC) are to be considered almost completely plastic as no appreciable deformation decreases have been observed with the progress of the excavation and subsequent processing.

The behavior observed in Sect. 3 appears fully compatible with the overpressure excavation methods (Beghoul and Demagh 2019) and with the very small distance between the excavation/processing boundary and the inclinometer tube (Table 4).

Table 4. Main parameters – Section 3

Tube length (m)	Measuring start	Measuring end	TBM transit	N° measures
42.5	05/10/2019	31/12/2019	08/11/2019	160

As part of the experimentation, considering the numerous measurements carried out even in almost undisturbed conditions, a series of in-depth analyses were carried out in the pre-excavation periods (Fig. 11a, 11b, 11c). Preliminary analyses in this area have confirmed the potential of the instrumentation in terms of accuracy and repeatability but have also posed a first series of questions on what are the real limits of the inclinometric measurement performed by a robotic system. The indications of literature (Mikkelsen 2003) and the rules of best practices, provide indications in terms of overall accuracy equal to about 7 mm per 30 m. The recent ISO standard (ISO 18674 2017) suggest for a 30 m tube length, ±2 mm on the main channel (plane with the wheels) and more than ±2 mm for the transversal channel. Some results discussed in this paper improve the standards for inclinometer measurement performed with a portable probe or with IPI. The reasons for these performances are mainly related to the robotic system approach, the performance of the acquisition algorithm and the execution of the double reading. Table 5 summarized some of the results obtained using all the raw measurements (about 330 cycles) without corrections and removing only 3% of anomalous data. A further synthesis attempt was made with the graph of Fig. 11d, showing the information acquired for the transverse and longitudinal channels in terms of the length of the tube. This graph shows that there is an inverse proportionality between the repeatability of the measurement and the length of the tube and that this is considerably more marked for the longitudinal channel (B). The illustrated values are, of course, also influenced by the characteristics of the inclinometer tube in terms of material, type of groove but above all diameter (Sappa et al. 1989). This parameter is a fundamental element both as regards the wheel-groove coupling and concerning the stress to which the centering/coupling springs of the inclinometer probe are subject. During the experimentation, the very small diameter

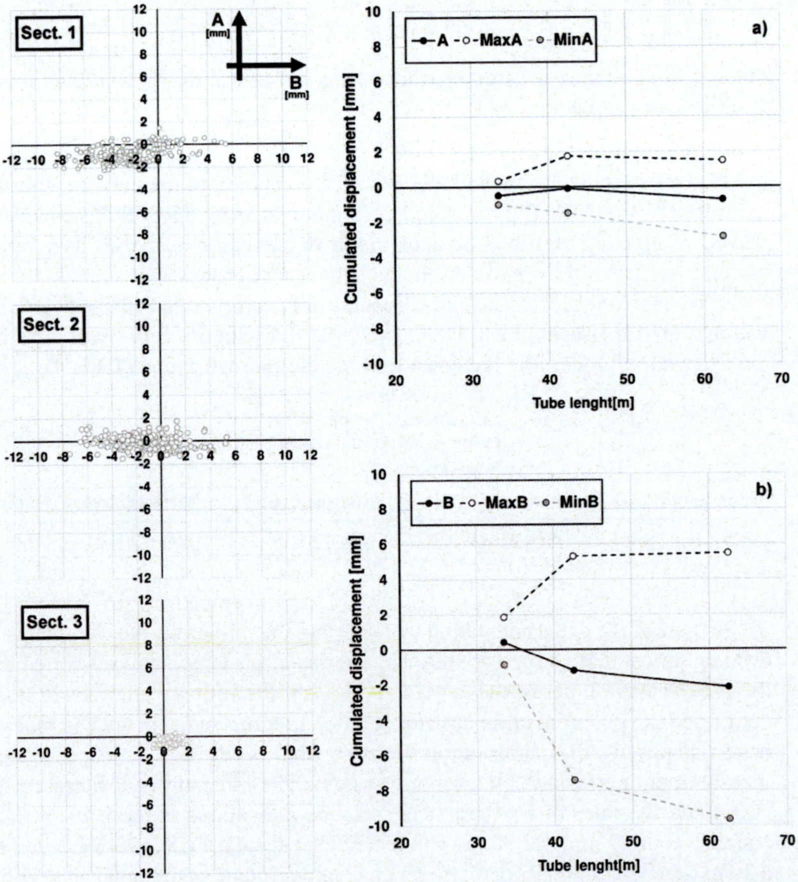

Fig. 11. Cumulated displacement at the top of the borehole in undisturbed periods in the three Sections (1, 2, 3). Subfigure a), b) show cumulated values at the top of the borehole as a function of tube length (mean, maximum and minimum values for each channel)

of the tubes significantly stressed the springs of the used probe causing two breakages, one for each wheel structure (upper and subsequently lower).

Table 5. Statistical analysis in the undisturbed phase, (cumulated displacements at the top of the borehole)

Sector.	Tube length (m)	Channel A	Channel B	N° measures
1	62.5	−0.7 ± 0.7	−2.2 ± 2.6	207
2	42.5	−0.1 ± 0.6	−1.2 ± 2.5	215
3	33.5	−0.5 ± 0.4	0.4 ± 0.6	95

4 Conclusions

The experimentation of the robotic inclinometric system in the context of large underground works can be considered successfully concluded. The instrumentation, previously used in landslides monitoring, has allowed to monitor three sections of the T3 section of the C line of the Rome metro with satisfactory, although preliminary, results. For all three sections, no deformations were detected that would induce problems on the priceless artistic and cultural heritage involved. In particular, for the first two, Basilica di Santo Stefano Rotondo al Celio and Via dei Fori Imperiali, no movements exceeding the instrumental and method tolerances were detected. As for the third section located at the Basilica di Massenzio, the probable combination of the site stratigraphy joint with the extreme proximity to the excavation section has produced a very limited but extremely clear and interesting deformation, from an engineering and geotechnical point of view. The possibility of making high frequency measurements (up to 6 measurements day) together with the accuracy of measurement and repositioning provided by a robotic system, has also allowed to follow up the evolution of the deformation over time with a vertical resolution rarely achievable with others systems. The execution times for the double measurement (0/180°) were between 50 min (section 3) and 1 h and 20 min (section 1) and were deemed compatible with the temporal evolution of the mechanized excavation.

The measurement activities, in addition to being functional for the monitoring of the excavation and for the designers, were also used to evaluate the measurement accuracy on three different sites as well as to evaluate the robustness and reliability of the system at frequencies of measurements that are unusual for the equipment. The main critical issues emerged are related to i) potential problems with the probe springs induced by inclinometer tubes with extremely small diameter ii) limitations in the positioning of the system with particular regard to the driveways open to vehicular traffic. These observations must be an integral part of a correct and effective design of the monitoring networks as well as a correct management strategy for the data collected.

Acknowledgments. The Authors kindly acknowledge Metro C S.c.p.a. and IMG S.r.l. companies for their cooperation and willingness.

References

Allasia, P., et al.: Deep displacements measured with a robotized inclinometer system. In: Proceedings of the 10th International Symposium on Field Measurements in Geomechanics, Rio de Janeiro, Brasil, July 2018 (2018)

Beghoul, M., Demagh, R.: Ground movements induced by tunnels excavation using pressurized tunnel boring machine-complete three-dimensional numerical approach. In: Badr, M., Lotfy, A. (eds.) Sustainable Tunneling and Underground Use, GeoMEast 2018. Sustainable Civil Infrastructures. Springer, Cham (2019)

Dunnicliff, J.: Geotechnical Instrumentation for Monitoring Field Performance, p. 577. Wiley, New York (1993)

Herrera, G., et al.: The differential slow moving dynamic of a complex landslide: multi-sensor monitoring. In: 4th World Landslide Forum, Ljubljana, Slovenia, 29 May–2 June 2017 (2017)

ISO 18674-3: Geotechnical investigation and testing - Geotechnical monitoring by field instrumentation - Part 3: Measurement of displacements across a line: inclinometers. ISO standard, Vernier, Geneva, Switzerland (2017). www.iso.org

Lollino, G., et al.: The use of the Automatic Inclinometric System (AIS) for landslide early warning: the case of Cabella Ligure (North -Western Italy). Phys. Chem. Earth **27**, 1545–1550 (2002)

Lollino, G., et al.: Time response of a landslide to meteorological events. Nat. Hazards Earth Syst. Sci. **6**(2), 179–184 (2006)

Mikkelsen, P.E.: Advances in inclinometer data analysis. In: Myrvoll, F. (ed.) Proceedings of the 6th International Symposium on Field Measurements in Geomechanics, Oslo, Norway, September 2003, pp. 555–566 (2003)

Sappa, M., et al.: Alcune considerazioni sulle misure inclinometriche. Notizie Associazione Nazionale Ingegneri Minerari 67–76 (1989)

Stark, T.D., et al.: Slope inclinometers for landslides. Landslides **5**, 339 (2008)

Seismic Site Response Characterization for Suez Canal Region, Egypt

Mohamed ElGabry[1(✉)], Hany M. Hassan[1], Franco Vaccari[1,2], Andrea Magrin[3], Fabio Romanelli[2], and Guiliano Panza[4,5,6,7]

[1] National Research Institute of Astronomy and Geophysics, Helwan, Cairo 11421, Egypt
elgabry@nriag.sci.eg
[2] Department of Mathematics and Geosciences, Via Weiss 4, 34127 Trieste, Italy
[3] OGS, Centro Ricerche Sismologiche, Istituto Nazionale di Oceanografia e di Geofisica Sperimentale, Via Treviso 55, 33100 Udine, Italy
[4] Institute of Geophysics, China Earthquake Administration, Beijing, China
[5] Beijing University of Civil Engineering and Architecture (BUCEA), Beijing, China
[6] International Seismic Safety Organization (ISSO), Arsita, Italy
[7] Accademia Nazionale dei Lincei, Rome, Italy
http://www.issoquake.org

Abstract. The Suez Canal Economic Zone (SCZone) is a mega project in Egypt that was launched to increase the role of the Suez Canal region in international trading and to develop the urbanization of the region. The reliable assessment of seismic hazard is a crucial input information for the designer and planner of all projects in this area. In this study, an updated seismic hazard map for Suez Canal region that incorporates revised historical earthquake catalogs, morphostructural zonation data (MZ), revised focal mechanism solutions and mechanical models of the lithospheric structure is provided. This is done within the framework of the Neo-Deterministic Seismic Hazard Assessment (NDSHA) procedure that may efficiently incorporate earthquake source rupture data, geological information and any reliable new information to adequately compute the earthquake ground motion maps, like PGA, PGV and PGD. This methodology provides a comprehensive source of ground motion records (acceleration, velocity, displacement) for Egypt where a substantial challenge is posed by the lack of recorded data. To consistently assess the uncertainty of our understanding of the seismic hazard and its effect on critical structures a sensitivity analysis is performed varying, for example, (a) source focal mechanism, (b) directivity (c) rupture process and (d) seismotectonic model. Starting from the initial definition of the source, bedrock and site properties based on the available knowledge of the characteristics of the different earthquake scenarios and of the soil structural properties of the area deduced from the ambient noise array measurements, a set of ground shaking scenarios have been computed at the selected profile crossing the Suez Canal. Variations of the ground shaking scenarios are evaluated looking basically at: (a) the amplitudes measured on the waveforms and the response spectra, (b) the changes in the shaking duration and (c) the modifications in the amplification patterns. Finally, the seismic input (e.g. response spectra, time histories) at the tunnel site that undergoes the Suez Canal to connect Sinai peninsula with the western side of the country is provided.

H. Shehata and M. Badr (Eds.): *Advancements in Geotechnical Engineering*, SUCI, pp. 59–78, 2021.
https://doi.org/10.1007/978-3-030-62908-3_5

1 Introduction

The Suez Canal Economic Zone (SCZone) is a mega project in Egypt that was launched on 5 August 2014. The project aims to increase the role of the Suez Canal region in international trading and to develop the three canal cities: Suez, Ismailia, and Port Said and to connect the western side of the country with the Sinai Peninsula. In addition, the project involves constructing a new city (new Ismailia city), an industrial zone, fish farms, completing the technology valley, expanding five existing ports, dredging a new canal parallel to the Suez Canal, and building two new tunnels between Suez Canal cities and Sinai (Fig. 1). The proper design of such strategic structures requires efficient and reliable computation of seismic hazard in order to provide the seismic input needed for the engineering studies for the structural assessment and non-linear behavior of the soil analysis.

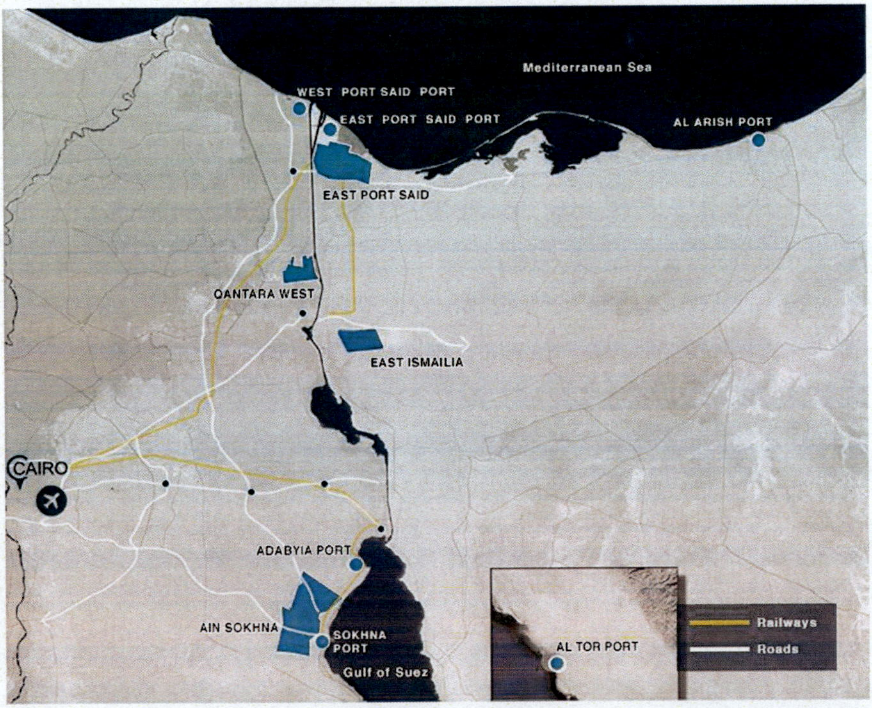

Fig. 1. The Suez Canal Special Economic Zone (SCZone) retrieved from https://www.sczone.e.g.

Seismic hazard assessment for a given site (site-specific seismic hazard) can be carried out using empirical or physics-based methods. Seismic hazard assessment that relies on physics-based models should be encouraged when dealing with infrastructure and strategic buildings. This method employs numerical modeling codes based upon the physical parameterization of fault rupture process and seismic wave propagation to reliably predict the ground motion parameters at different geographic scales and levels of details.

In this study physically based method following the Neo-deterministic seismic hazard assessment (NDSHA) method (Fasan 2017) is used. In this paper the Maximum Credible Seismic Input (MCSI) is calculated at the site of new Ismailia that encounter anew city as well as 2 tunnels crossing below the Suez Canal. The seismic input is calculated at two levels of details both to be used as reliable seismic input by the structural and geotechnical engineers. The first level of detail provides the "Maximum Credible Seismic Input at bedrock" ($MCSI_{BD}$), deliberately ignoring the "so-called" site-effects that, due to the tensor nature of earthquake ground motions, are not persistent but strongly dependent on the source process, and the transmitting (pathway) medium (Molchan et al. 2011; Hassan et al. 2017a). At this level, the same input data used in Hassan et al. (2017b) is used to compute the MCSIBD at the site. In the second level of detail, the Regional Scale Analysis (RSA) is used as a reference to select the most dangerous sources for the site and ground motion parameter of interest. As for these sources, a detailed Site-Specific Analysis (SSA), which considers the local structural properties (Site effect), is then carried out for each pathway as described in the next section. The SSA is a prerequisite to determine the "Maximum Credible Seismic Site-Specific Input" ($MCSI_{SS}$).

1.1 Geology and Seismicity of the Study Area

The geology of El Ismailia area is dominated by a sedimentary succession ranging from Eocene to Quaternary, with Mid-Tertiary basalts. The geological units are shown as groups, formations and smaller lithological units according to the standard international and local stratigraphic terminology. Geologic mapping in various scales and studies have been carried out previously in the investigated area by many authors (e.g., Barakat and About Ela 1970; El Shazly et al. 1974).

Information about subsurface geology is available only for formations younger than Eocene. Eocene is made of carbonate facies of 400 to 500 m thick mostly occupied the structural tableland, and it is overlain by sands and gravels (maximum thickness 80 m) then capped by Mid-Tertiary. It is mostly exposed along the foot-hill slopes of the Cairo-Gebel Shubrawit structural depressions. In the subsurface, the Oligocene sediments extend and generally slope northwards. The Miocene at the subsurface of the study area is represented mainly by the marine sandy limestones with depth up to 140 m. Most of the faulted structural ridges are built up of these facies. It is expected that the early Pliocene sediments are of fluviomarine clays and sandy clays. They represent an aquiclude horizon for the upper layers (Fig. 2). The surface geology map shows that the site of Ismailia tunnel is covered with Nile silt and clay and unconsolidated gravel and sand which make the area susceptible to liquefaction hazards.

At El Qantara area, the thickness of the Early Pleistocene sediments reaches 70 to 80 m. However, no detailed studies are available for them at present. Late Pleistocene-Holocene sediments cover the area due north of Khabret Umm Gidam and the Isthmus stretch. They are characterized by the alternations of gravelly sands, sands, slightly calcareous sandstones and clays. They attain a thickness ranging from 10 to 70 m, overlying the early Pleistocene, with thickness increasing toward the north. Holocene sediments are represented by sand sheets particularly along the eastern side of the Isthmus

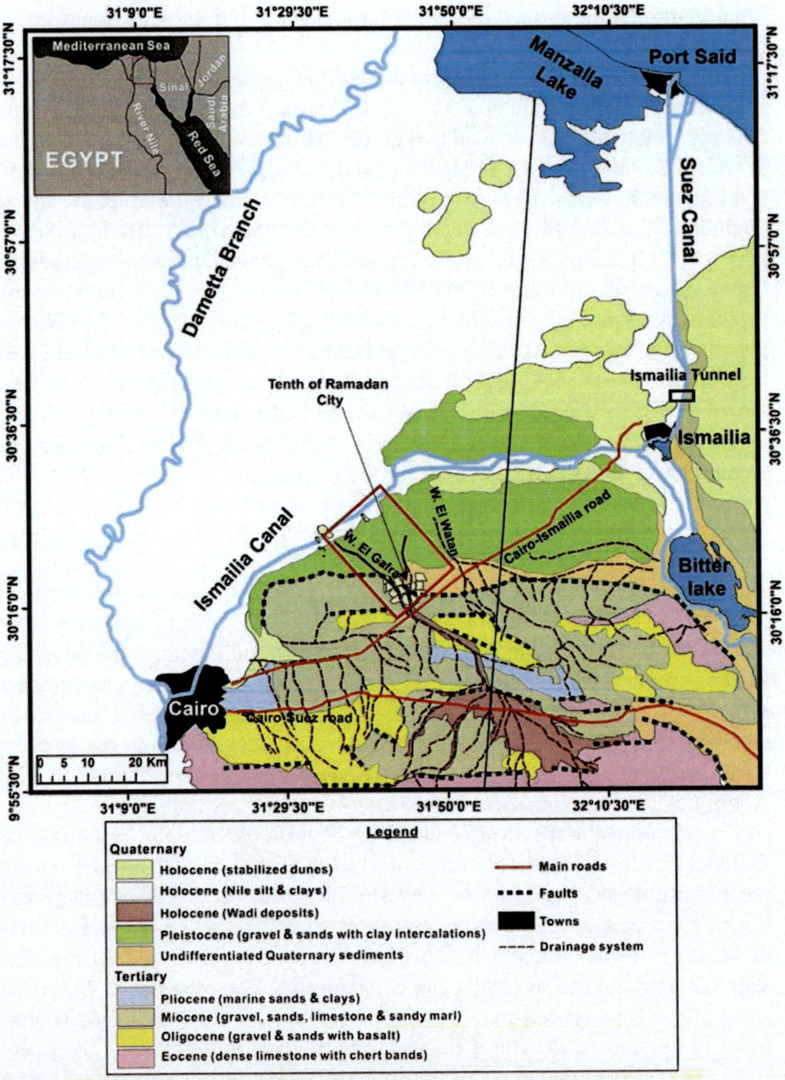

Fig. 2. Geologic map of the eastern Nile Delta on which the site of the tunnel is shown (Modified after Khalil et al. 2015)

stretch, and sand sheets and hammocks west of El Tell El Kabir-El Salhiya plain. Sabkhas and salt marshes dominate the northern strip of El Tell El Kabir-El Salhiya plain.

According Said (1990) to Ismailia and Bitter Lakes area is affected by two major active seismic trends; Northern Red Sea-Gulf of Suez-Cairo-Alexandria-Clysmic-trend and East Mediterranean-Cairo-Fayum Pelusiac trend. These two active seismic trends are genetically related to the well-known tectonic trends in Egypt.

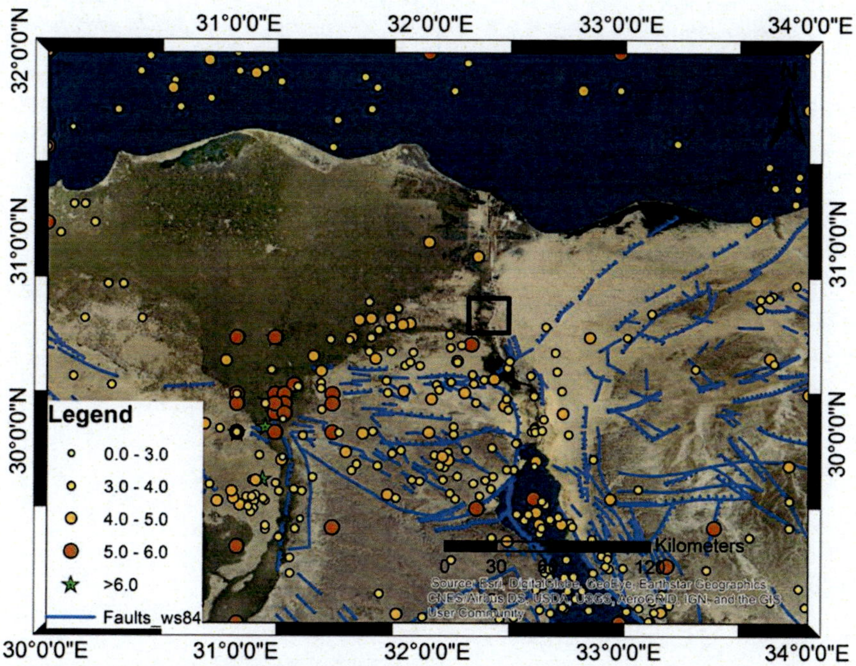

Fig. 3. Seismicity of the study area. Black square defines the site of the tunnel. Blue lines are geologic faults defined by EGSMA (1981)

Seismicity of the Suez Canal region generally and Ismailia particularly has been studied by very few studies (e.g., Hegazi et al. 2013). The seismicity map of the study area shows a sparse seismic activity around the city, along the Cairo-Suez district and Gulf of Suez rift (Fig. 5). The morphostructural zonation work done by Gorshkov et al. (2019) has revealed a number of seismogenic nodes, which are capable of generating earthquakes of magnitude 5+ near the study site marked by a black square in Fig. 3. The new data should be incorporated in any seismic hazard assessment particularly for strategic and critical structures in order to ensure the safety of such critical structures.

2 Methodology

The Neo-Deterministic Seismic Hazard Assessment (NDSHA) (Panza et al. 2001, 2012 and its update in Magrin et al. 2016) is a scenario-based procedure which supplies realistic time history ground motions, which are calculated as the tensor product of the earthquake source and the Green's function of the medium, thus preserving the tensor nature of the physical problem (Fig. 4). NDSHA computation is based on the maximum magnitudes occurred within sources that may affect the studied site regardless of their rate of occurrence. Very realistic synthetic seismograms that represent the knowledge of the earthquake rupture process and of the inelastic pathway mechanical properties. The computed seismograms supply reliable estimates of engineering relevant parameters

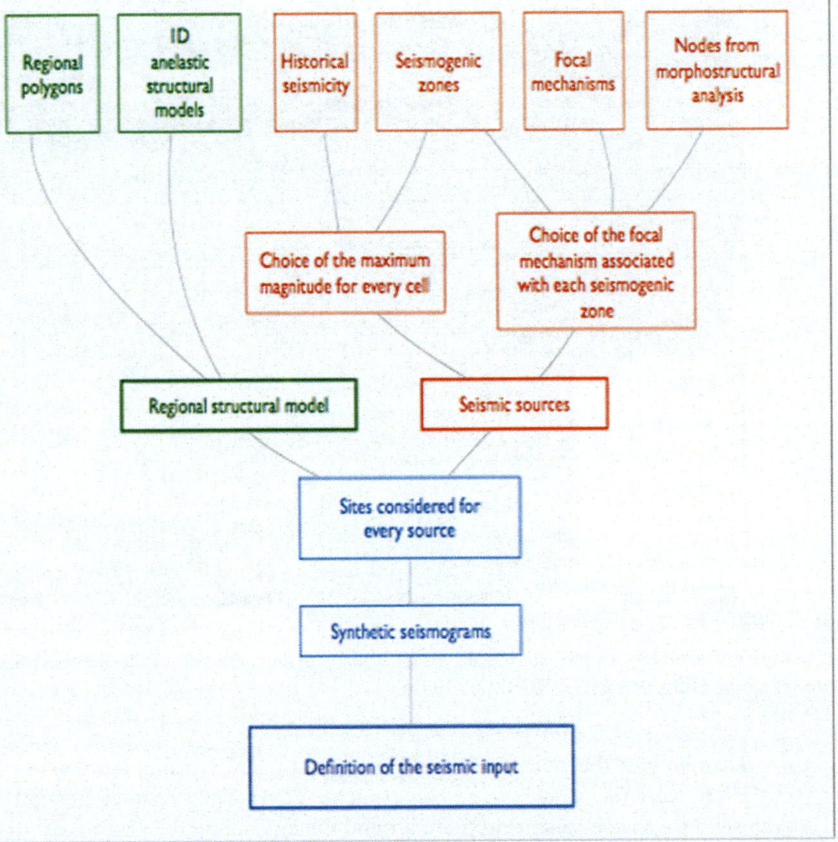

Fig. 4. Flow chart of the different steps in the NDSHA approach for the regional scale analysis.

such as peak ground motion parameters and the maximum credible seismic input (MCSI) (Fasan et al. 2015; Fasan 2017). The seismograms computed at the site of interest can be used immediately as input for the Non-Linear Time History Analysis of any built structure without any further processing (e.g., filtering or scaling). In fact, realistic time histories are of a great importance for structural and technical engineers willing to design a new structure and/or evaluate the seismic performance of the existing built environment and to investigate the non-linear behavior of soil at the site of interest.

At the national scale, NDSHA makes use of information about the spatial distribution of large magnitude earthquakes, as defined by seismic history and seismotectonics and an extensive set of relevant geological and geophysical data (Zuccolo et al. 2011). The basic product of this method is a map in which the maximum of a given ground motion parameter - peak ground acceleration (PGA), velocity (PGV) and displacement (PGD) - or any other parameter relevant to seismic engineering, which can be extracted from the computed seismograms, is associated to each site.

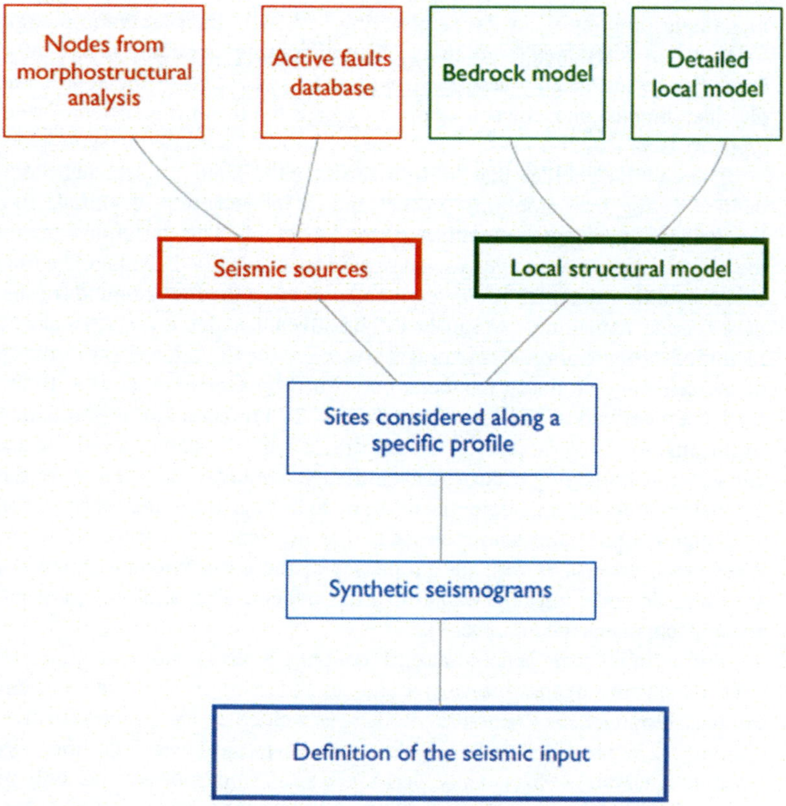

Fig. 5. Flow chart of the different steps in the NDSHA approach for the site scale analysis.

The NDSHA procedure is very flexible and can be enhanced easily: the computer codes are constantly improved by incorporating all relevant progress in the knowledge of the physics of geological processes and in their numerical modeling.

2.1 NDSHA Procedures

Each seismogenic source in NDSHA at regional scale is arbitrarily placed at the center of a 0.2° × 0.2° cell of a regular grid that covers the study region, and which is called cellular source. Each cellular source is modeled as a scaled point-source and is characterized by focal mechanism and earthquake magnitude. Cellular sources are defined taking into account the available information, as provided by the seismotectonic model, the morphostructural zonation and reported seismicity. Namely, the location of possible future earthquakes is constrained by the seismogenic zones and by the seismogenic nodes. In the first step of cellular sources definition (discretization), earthquake epicenters reported in the catalog are grouped into 0.2° × 0.2° cells, and to each cell, the maximum magnitude recorded within it is assigned. A smoothing window is then applied, so that earthquake magnitudes are copied in the neighboring cells, and the maximum magnitude source for each cell is retained (smoothing). The smoothing procedure is basically needed for handling the uncertainty associated with the localization of the source and the linear physical dimension of the source often exceeds the linear dimension of a cell (which is typically 0.2° × 0.2°).

The cellular sources that lie in a seismogenic zone are selected among the ones defined during the smoothing process and, if the resulting magnitude in each cell is lower than 5, a magnitude five is assigned by default. This selection is mainly based on the hypothesis that, wherever a seismogenic zone is defined, damaging earthquakes may occur, and the value of 5 is conventionally taken as the lower bound for the magnitude of moderate damaging earthquakes. Seismogenic nodes are earthquake- prone areas identified through morphostructural analysis. For the Egyptian territory, the identification of the seismogenic nodes has been performed by Gorshkov et al. (2019). The nodes have been defined as circles of radius R = 20 km centered at points of intersection of lineaments. Such dimension is comparable with the size of a seismic source, which can generate damaging events. A magnitude threshold of 5 has been considered for the estimation of the seismic potential of the node through the pattern recognition technique. A double-couple point source is located at the center of each cell, assigning a focal mechanism consistent with the properties of the corresponding seismogenic zone or seismogenic node. If a cell belongs to a seismogenic zone and a seismogenic node, both focal mechanisms are considered.

For regional-scale seismic hazard mapping, a "Regional Scale Analysis" (RSA) is carried out using all possible seismic sources and a simplified sequel of different adjacent laterally homogeneous structural models, in welded contact, representative of bedrock conditions. When a detailed analysis is needed, to duly model the site-effects, a "Site-Specific Analysis" (SSA) can be performed (Fig. 4) introducing not only the modeling of structural and topographical heterogeneities along a specific profile but also of the influence of the source rupture process on the seismic wavefield at a site. So far, the NDSHA method has been applied in several countries at different levels of detail (e.g., Mourabit et al. 2014; Panza et al. 2012; Hassan et al. 2017b; El-Sayed et al. 2004; Lokmer et al. 2002).

3 Computation of Seismic Input

3.1 Computation of $MCSI_{BD}$

The sources criteria and structural models of the Earth needed in order to perform MCSI are taken from the study of Hassan et al. (2017b). The NDSHA-MCSI accommodates the available, reliable information about the spatial distributions of the sources, their magnitudes, and focal mechanisms, as well as about the mechanical properties of the inelastic media along the transmitting pathway.

Actually, for MCSI computations dedicated to reliable engineering analyses, MCSI should be obtained using a significant number of realizations. For example, a number of 100 realizations of the rupturing process is used, which should be enough for most applications.

In this work, three-component synthetic seismograms are computed at the free surface at the new Ismailia site. The cutoff frequency is 10 Hz in the domains of displacement, velocity, and acceleration. All the potential sources, within an epicentral distance of about 300 km, have been considered taking into account. To accommodate uncertainty in all input parameters; parametric test has been performed considering 100 different realizations of the source rupturing process, for three directivity angles (i.e., 0°, 90° and 180°) of two different faulting styles (unilateral and bilateral). The focal mechanisms have been adopted based on the available earthquake mechanism solutions for the study area. The MCSI has been computed by joining the statistics for all the computed scenarios for the tunnel site.

Figure 6 shows the MCSI spectrum computed at the bed rock at the site of the tunnel. The individual spectra indicate that three earthquake scenarios are controlling the spectral acceleration. Two are dominating the short period part of the RS: a seismogenic node with M = 5.5 and an instrumental earthquake that occurred in 1987 with M = 5. The long period part is controlled by a distant earthquake scenario: the 950 earthquakes near Cairo. At this level, the MCSI represents a sort of Uniform Hazard Spectra (UHS) (Trifunac 2012), where the hazard is identified by the maximum magnitude expected for every potential source that could affect the sites of interest. The earthquake scenarios defined at this level are used for the detailed site-specific analysis described in the next section, using the seismogenic node as a near source and the historical event as a distant source.

3.2 Computation of $MCSI_{SSA}$

Based on the information described in previous section about the earthquake scenarios that control the seismic hazard at the site of the tunnel, a SSA has been performed. A laterally heterogeneous profile representative of the local conditions has been composed from Maklad et al. (2020).

The earthquake scenario used for the SSA has been chosen from the disaggregation of the MCSIBD response spectrum calculated with the RSA (i.e., the source-distance combination that gives the largest spectral acceleration at periods of interest) in order to optimize computational time. The controlling events, for the range of periods from 0.0 s to 4.0 s, have been found to be magnitudes 6.5–7.0 (950 earthquakes with M = 6.5–7.0) and 5.5 (seismogenic node with M = 5.5) earthquake scenarios at an epicentral distance in the range from 20 to 100 km (Fig. 7).

A hybrid technique (see Fäh et al. 1993; Panza et al. 2001), based on modal summation (MS) and finite difference (FD), is used to simulate the earthquake ground motion at the selected sites. The computation is made considering three earthquake scenarios, two examples, a distant earthquake scenario with Magnitude 6.5–7.0 and a seismogenic node scenario with M = 5.5 are presented. The technique allows to handle the seismogenic characteristics of the source and the properties along the pathway with full respect of the tensor nature of earthquake ground motion (Aki and Richards 1980). Three-component synthetic seismograms are computed on the free surface along the considered profiles, with frequency content as high as 5 Hz in the domains of displacement, velocity, and acceleration. Two source models have been considered, taking into account changes in magnitude, rupture directivity, and mechanism. Parametric tests have been carried on

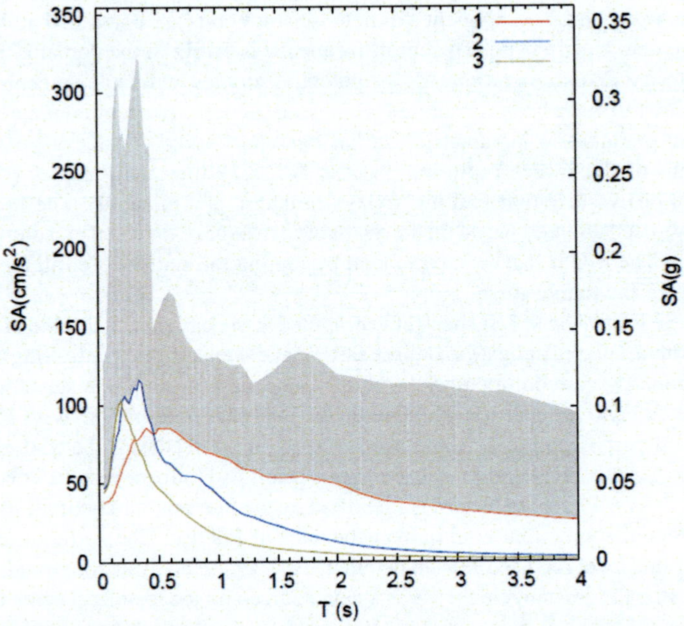

	source	profile	M_w	edi (km)	depth (km)	strike (°)	dip (°)	rake (°)	sre (°)	slon (°N)	slat (°E)
1	sz 0007	-	7.2	42.6	15.0	248	80	190	175	31.9	30.5
2	nn 1010	-	5.5	13.1	10.0	248	80	190	238	32.3	30.5
3	sz 0007	-	5.0	9.5	10.0	248	80	190	81	32.3	30.7

Fig. 6. MCSIBD (grey areas correspond to the values between median and 95th percentile) for the study site.

varying the characteristic of the scenario earthquakes (e.g., focal mechanism parameters). Response spectra (RS) have been computed starting from the synthetic accelerograms, and amplifications have been estimated in terms of response spectra ratios (RSR), using the RS computed for a bedrock model as a reference.

3.3 Model Parameters

In order to perform reliable synthetic seismogram computation, one has to properly model the characteristics of the source generating the seismic energy, and of the transmitting (pathway) medium.

A) Source

Synthetic seismograms are computed in the scaled point-source approximation, using a double-couple of forces placed at the hypocenter defined accordingly with the known

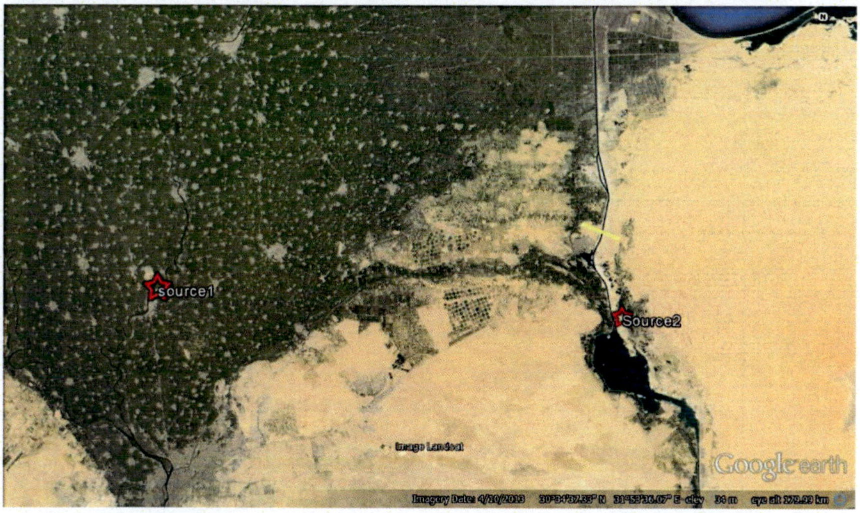

Fig. 7. Earthquake scenarios and the target site.

focal mechanism properties, hypocentral depth, and magnitude. Directivity effects, different rupture styles, and rupture realizations are also taken into account. Several parameters can be adjusted when generating the Source Time Functions (STF) according to the methodology proposed by Gusev (2011). Two earthquake scenarios are defined (see Fig. 7):

1. Historical event 935 A.D., located near the Nile Delta, M = 6.5–7.0. Distance to the target area about 100 km.
2. Morphostructural zonation node, located near Ismailia, M = 5.5. Distance to the target area about 20 km.

B) Reference bedrock model

The seismic energy radiated from the source propagates in the bedrock model towards the local profile. The bedrock model is representative of the average structural properties between the source and the site of interest; the bedrock model is also later used as a reference model for the estimation of the amplifications along the considered profile.
The bedrock models adopted in this study are the ones that have been recently adopted for NDSHA computations at a national scale (Hassan et al. 2017b), embedding the target profile for two different scenarios (see Fig. 8). The layerings of the two complete bedrock models are shown in Fig. 8.

C) Local Site model

This is the part of the structure corresponding to the selected sites (see Fig. 9). The local model has been set up after Maklad et al. 2020 using the data from noise vibration measurements carried out along the profile and from borehole data dug near the study site, and it is shown in Fig. 9. The heterogeneous part of the local structure is 10 km long and 300 m thick. Given the properties of the slowest layer considered in the local

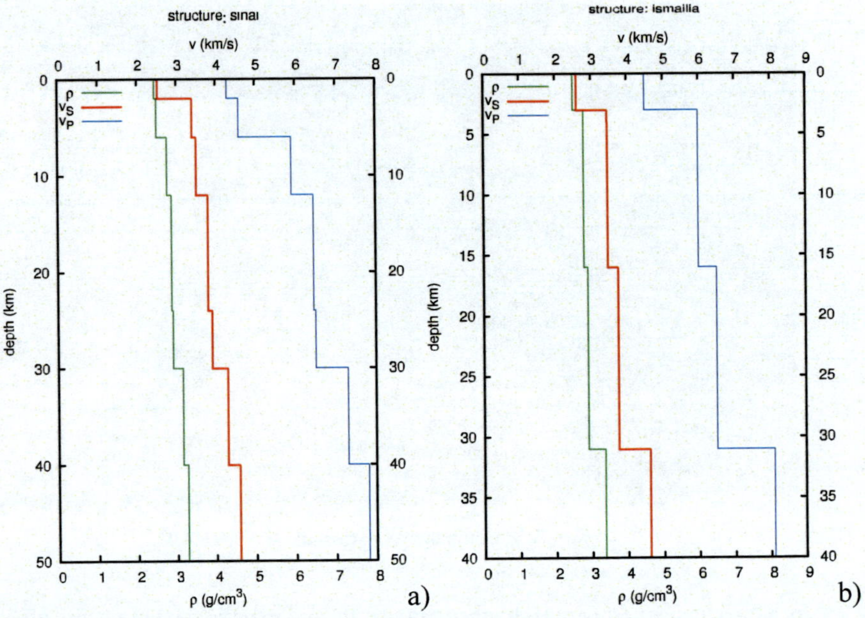

Fig. 8. Reference bedrock model Sinai (a, for the eastern zone) and Ismailia (b, for the western zone).

structure (260 m/s), and the highest frequency contained in the spectrum of the synthetic seismograms, the adopted grid step in this part of the model is 5 m. The FD mesh is continued at depth adopting the properties of the reference bedrock, down to a depth larger than the source depth. The grid step in this deeper and faster part of the model is 30 m. Figure 9 shows the path of the profile (top), the resulting 2-D model (middle) and the mechanical properties of the 2-D model layers together with the location of two sites selected in order to compute the seismic input at both ends of the tunnel.

4 Ground Shaking Scenarios

Once defined the properties of the source, of the bedrock and of the local structure, a preliminary test is required, to confirm a) that the numerical model has been duly prepared, b) that the incoming wavefield is properly introduced in the FD mesh and c) that the unwanted reflections from the artificial boundaries of the FD model are properly removed.

The test is made creating a hybrid model where the characteristics of the bedrock are assigned also to the properties of the local structure where the FD technique is applied. Therefore, a laterally homogeneous structural model is used. This model can be treated both with the modal summation and the finite difference techniques. The seismograms obtained with the two methods are compared, and their adequate similarity confirms that the finite difference mesh has been properly constructed.

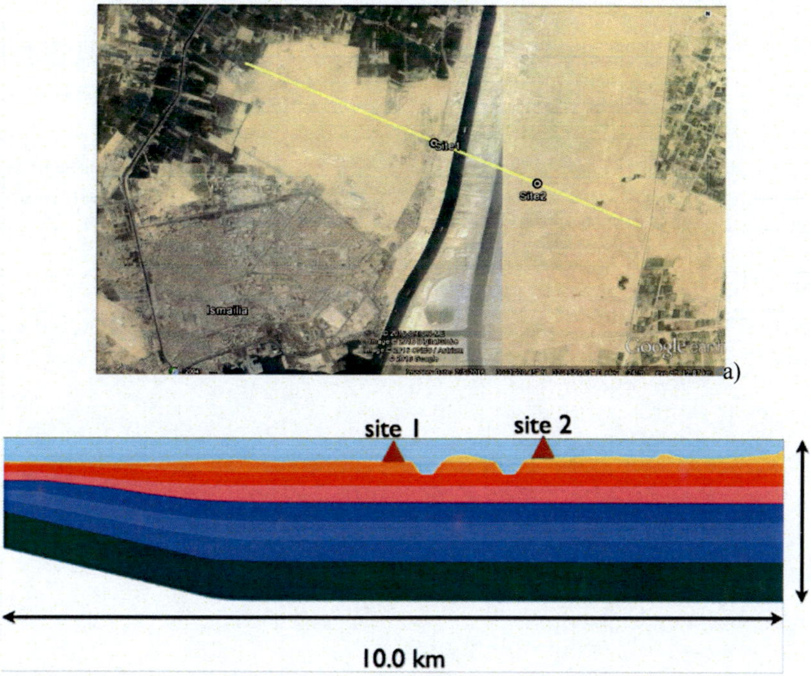

b)

Descr.	Rho	Vp	Qp	Vs	Qs
Sand 1	1.000	0.450	44	0.260	20
Sand 2	1.200	0.540	44	0.310	20
Sand 3	1.300	0.610	44	0.350	20
Sand 4	1.500	0.700	44	0.370	20
Sand and Shale 1	1.600	0.800	110	0.450	50
Sand and Shale 2	1.700	0.950	110	0.550	50
Sand and Shale 3	1.750	1.100	110	0.640	50
Sand and Shale 4	1.800	1.500	110	0.900	50

c)

Fig. 9. Target profile along the Suez Canal (a), with its section (b) and the properties of the seismological units.

4.1 Ground Shaking for the Local Model: Scenario 1

Synthetic seismograms and amplification maps for the local model and Scenario 1 are shown in Figs. 10 and 11, respectively. The three-component synthetic seismograms in terms of displacement, velocity and acceleration have been computed with the hybrid method for the historical event 935 A.D. (Scenario 1) for an array of about 100 sites regularly spaced every 100 m along the EW profile that across the Suez Cana and represent the path for the Ismailia tunnels (Fig. 10). The peak ground displacement, velocity and acceleration of 1.68 cm, 7.0 cm/s and 39.0 cm/s^2. The response spectra ratio (RSR) is other parameters crucial for engineering purposes. The distribution of RSR as a function of frequency and the distance from source is producing the amplification patterns. A maximum amplification of 8 at frequency less than 2 Hz was estimated for the radial component from scenario 1.

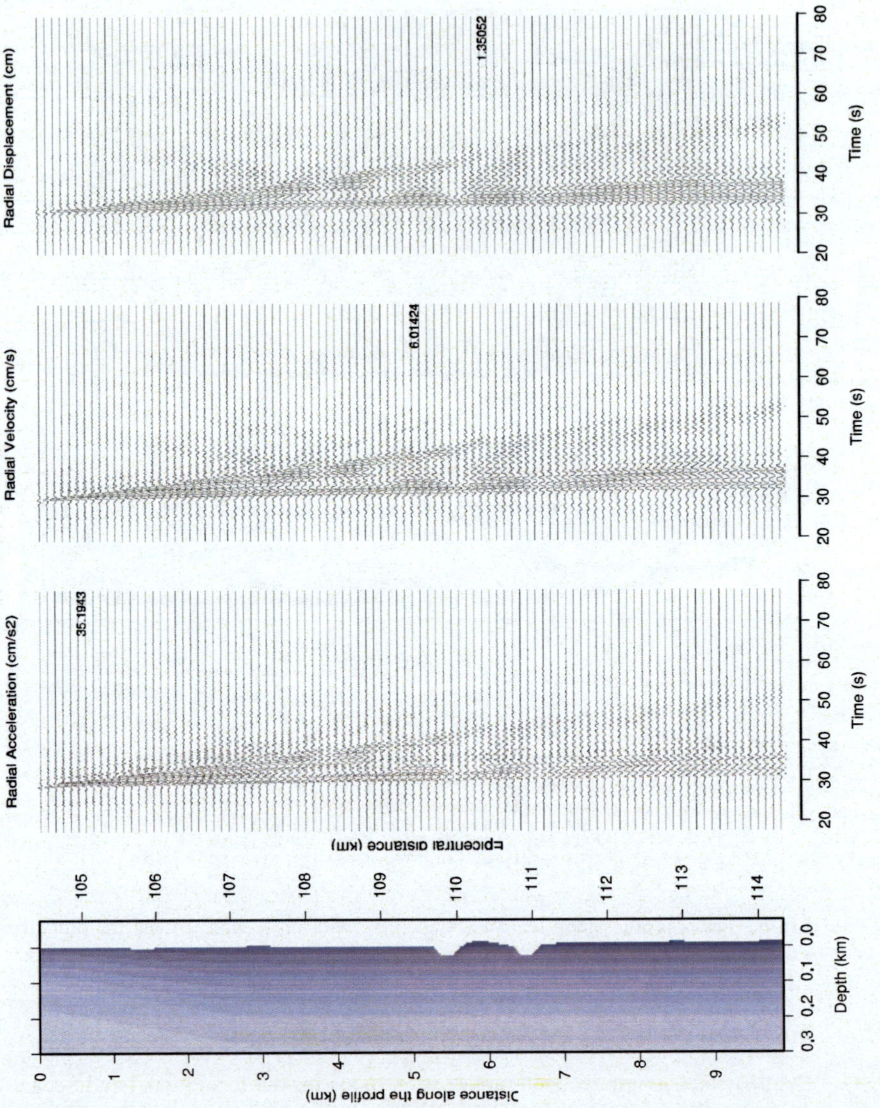

Fig. 10. Radial component synthetic seismograms along the profile. Accelerations in cm/s^2, velocities in cm/s, displacements in cm.

4.2 Ground Shaking for the Local Model: Scenario 2

Synthetic seismograms and amplification maps for the local model and Scenario 2 are shown in Figs. 12 and 13, respectively. The synthetic seismograms have been computed for the same profile but different earthquake scenario (Scenario 2), which is a Morphostructural node delineated by the Gorshkov et al. (2019). The Figs. 12 show

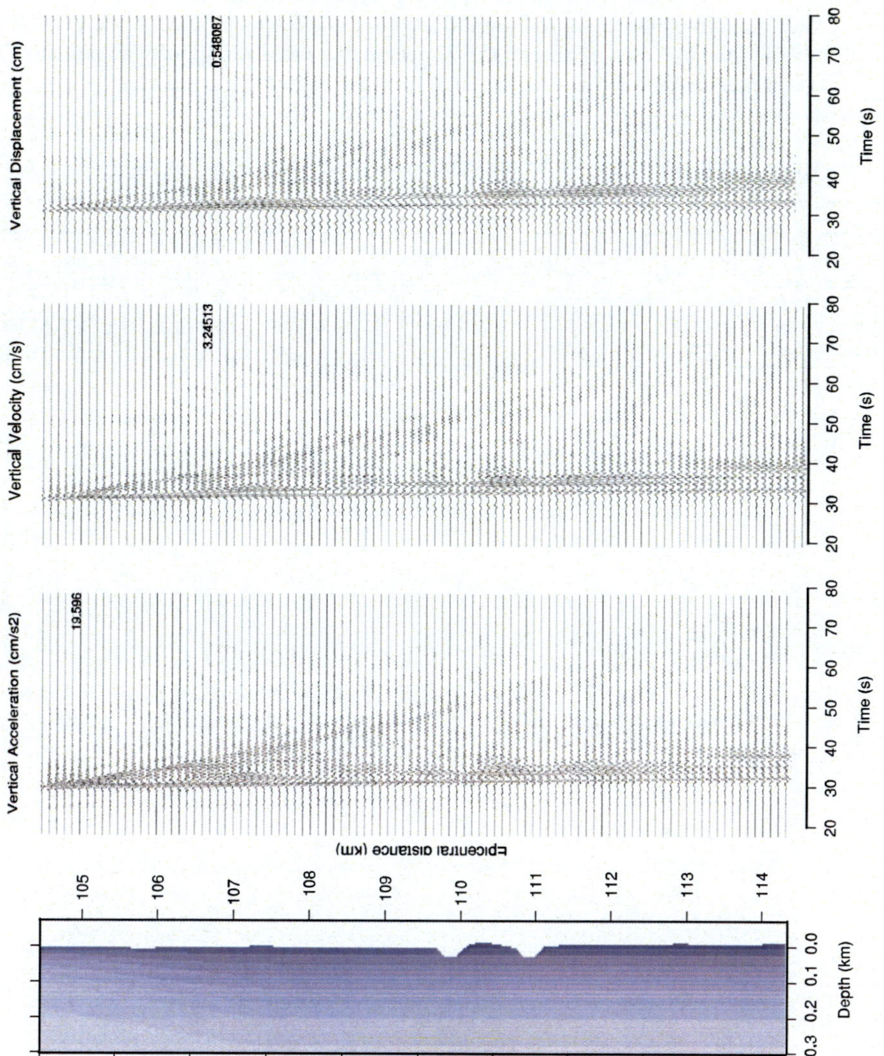

Fig. 11. Vertical component synthetic seismograms along profile. Accelerations in cm/s^2, velocities in cm/s, displacements in cm.

maximum ground motion parameters, i.e. displacement, velocity and acceleration, are 1.0 cm, 10.1 cm/m and 205.7 cm/s^2, respectively. The RSR along the profile were computed up to frequency of 5 Hz. The maximum amplification is obtained for the radial component of about 7 at lower frequency less than 2 Hz. Higher values of amplification are concentrated to in the western side at shorter epicentral distance where as on the eastern side of the canal the amplification is much less. Although the peak acceleration is observed on the transverse component, the radial component exhibits the highest amplification factors.

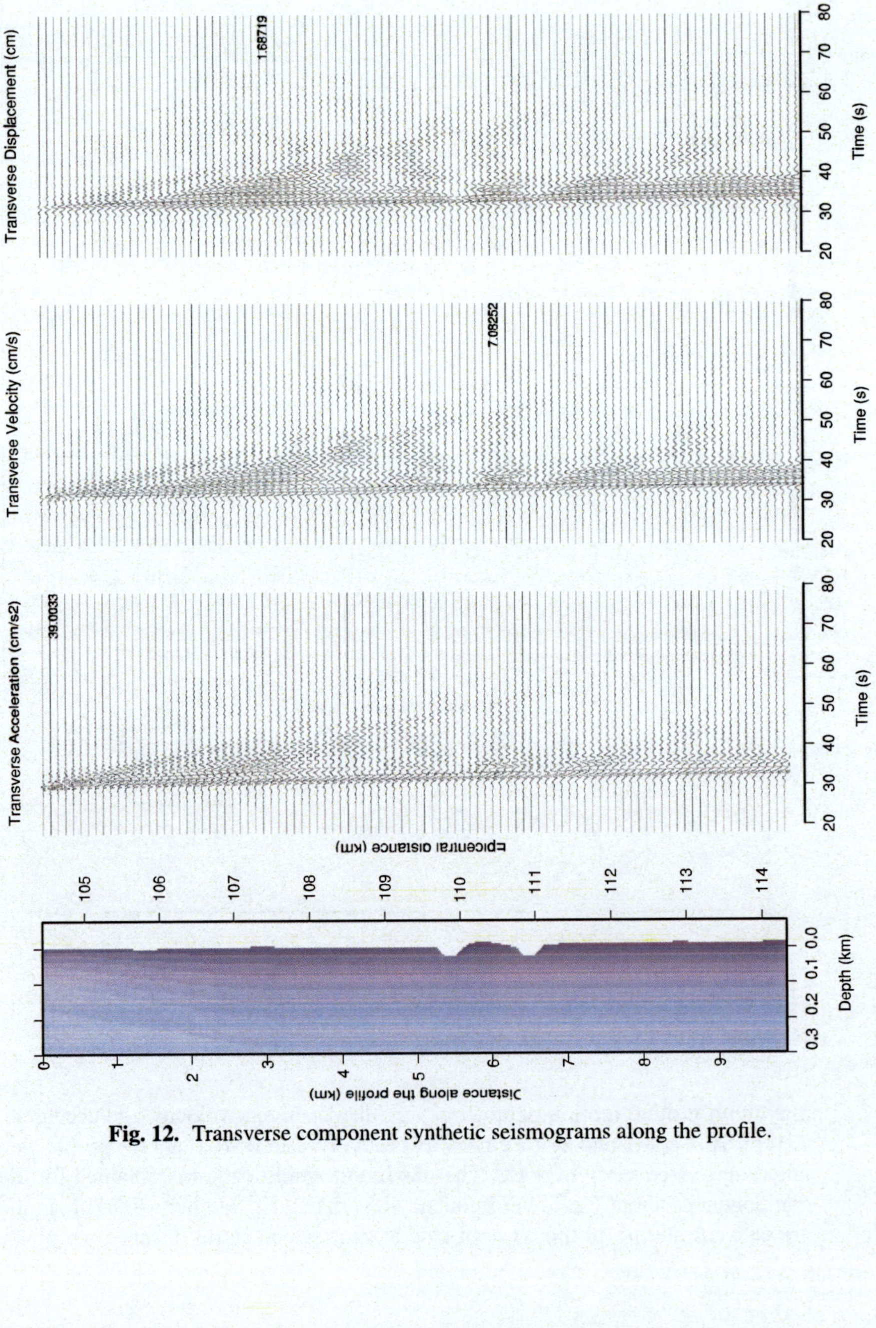

Fig. 12. Transverse component synthetic seismograms along the profile.

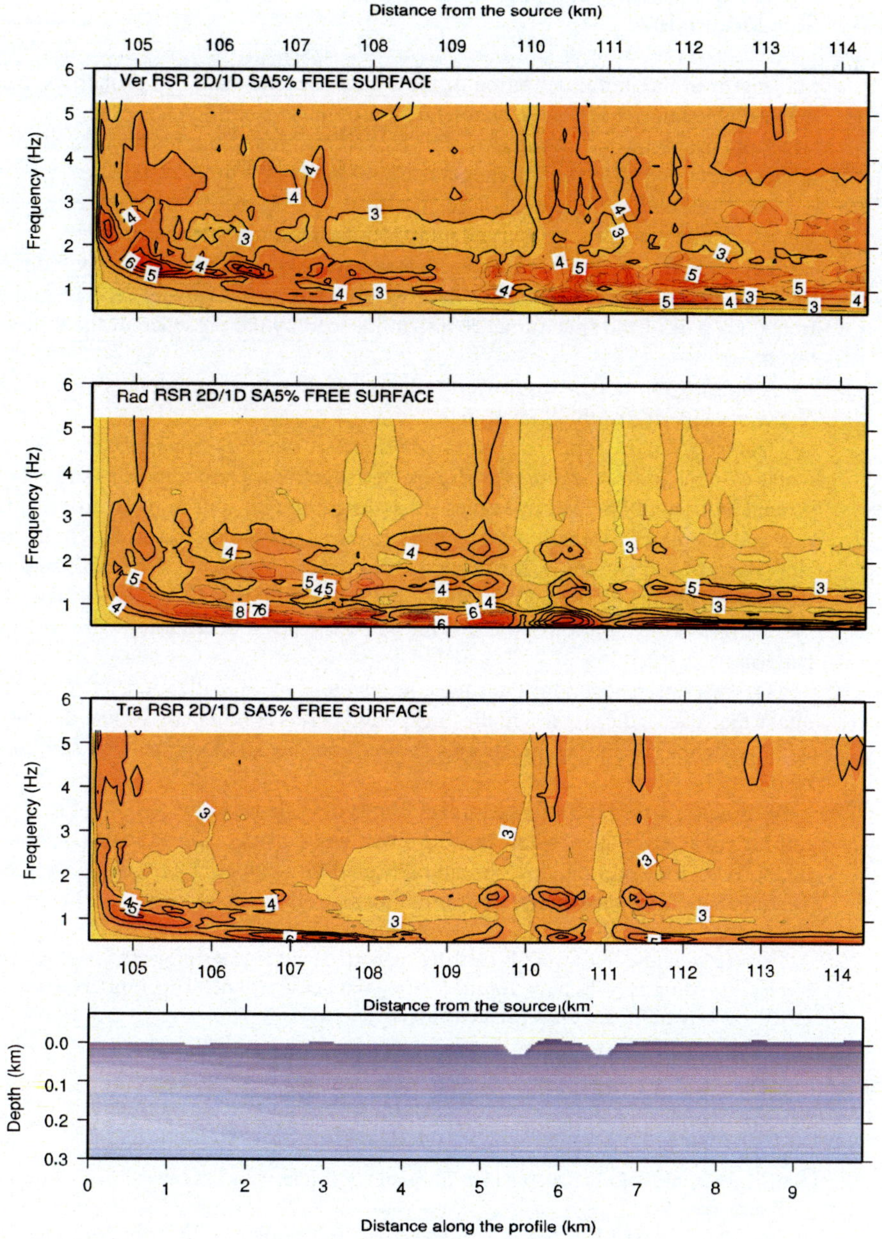

Fig. 13. Amplification pattern for the profile (RSR/Ismailia).

5 Conclusions

Starting from the initial definition of the source, bedrock and site properties, based on the available knowledge of the characteristics of the two earthquake scenarios and of the structural properties of the area, a set of ground shaking scenarios have been computed at the selected profile crossing the Suez Canal in the proximity of Ismalia.

The site condition has a significant effect on the amplification of seismic signals where amplification were observed up to 8 times the original signal.

The amplification is generated by the presence of the thin low-velocity sediments on top of the model. Velocities and displacements, having relatively less energy in the high-frequency part of the spectrum, are less influenced by the presence of the thin topmost layers.

Amplification in the western side where the old city of Ismaliya is higher than that on the eastern side of the Suez Canal.

The calculated Response spectra could reflect that it is impossible to use only the worst-case scenario to account for all expected spectrum. The complex nature of expected ground motion will not survive unless including different sources with different realization and keeping the tensor nature of the problem that holds the directivity nature and site effect. Modifications of the ground shaking scenarios have to be evaluated looking basically a) at the amplitudes measured on the waveforms and the response spectra, b) at the changes in the duration of the shaking, and c) at the modifications to the amplification patterns.

The response spectra for the site is the envelope of all realizations of scenario earthquake that affects the site where the long period part is dominated by more distant, events with larger magnitude, while periods shorter than 1 s are dominated by smaller events nearer to the site.

Physically based hazard assessment brings the balance in the pursuit of realistic seismic hazard as in the areas where lacking ground motion records and sparse seismicity; the method could accommodate information from different techniques including; Geologic, Tectonic, and Morphostructural information as well as account for their uncertainty.

Acknowledgments. This work is partially supported by the ASRT project "Suez Canal regional Center for Mitigating Natural Hazard". The Authors are grateful to Prof. Hesham Hussein (NRIAG) for fruitful discussion and continuing guidance.

References

Aki, K., Richards, P.G.: Quantitative Seismology: Theory and Methods. W.H. Freeman, San Francisco (1980)

Barakat, M.G., About Ela, N.M.: Microfacies and paleoecology of Middle Eocene and younger sediments in Geneifa area, Cairo-Suez district. UAR J. Geol. **14**, 23 (1970)

El-Sayed, A., Vaccari, F., Panza, G.F.: The Nile Valley of Egypt: a major active graben that magnifies seismic waves. Pure Appl. Geophys. **161**(5), 983–1002 (2004)

EGSMA (Egyptian Geological Survey and Mining Authority): Geologic map of Egypt 1:2000000. Ministry of Industry and Mineral Resources, Cairo, Egypt (1981)

El Gezeery, M.N., Marzouk, I.M.: Miocene rock-stratigraphy of Egypt, Stratigraphic Subcommittee of the National Committee of Geological Sciences, Egypt. J. Geol. **18**, 1 (1974)

El-Shazly, E.M., A'adel Hady, M.A. El Ghawaby, M.A., El Kassas, I.A., El Shazly, M.M.: Geology of Sinai Peninsula from ERTS-1 satellite images. Remote Sensing Research Project, Academy of Scientific Research and Technology, Cairo, Egypt (1974)

Fäh, D., Iodice, C., Suhadolc, P., Panza, G.F.: A new method for the realistic estimation of seismic ground motion in megacities: the case of Rome. Earthq. Spectra **9**(4), 643–668 (1993)

Fasan, M., Amadio, C., Noè, S., Panza, G., Magrin, A., Romanelli, F., Vaccari, F.: A new design strategy based on a deterministic definition of the seismic input to overcome the limits of design procedures based on probabilistic approaches. In: XVI ANIDIS Conference, L'Aquila, Italy. arXiv preprint arXiv:1509.09119 (2015)

Fasan, M.: Advanced seismological and engineering analysis for structural seismic design. Ph.D. thesis, Trieste University – Italy (2017)

Gorshkov, A.I., Hassan, H.M., Novikova, O.V.: Seismogenic nodes ($M \geq 5.0$) in Northeast Egypt and implications for seismic hazard assessment. Pure Appl. Geophys. **176**(2), 593–610 (2019)

Gusev, A.A.: Broadband kinematic stochastic simulation of an earthquake source: a refined procedure for application in seismic hazard studies. Pure. Appl. Geophys. **168**, 155–200 (2011). https://doi.org/10.1007/s00024-010-0156-3

Hegazi, A.M., Seleem, T.A., Aboulela, H.A.: The spatial and genetic relation between seismicity and tectonic trends, the Bitter Lakes area, North-East Egypt. Geoinf. Geostat.: Overview **1**(2), 7 (2013)

Hassan, H.M., Panza, G.F., Romanelli, F., ElGabry, M.N.: Insight on seismic hazard studies for Egypt. Eng. Geol. **220**, 99–109 (2017a)

Hassan, H.M., Romanelli, F., Panza, G.F., ElGabry, M.N., Magrin, A.: Update and sensitivity analysis of the neo-deterministic seismic hazard assessment for Egypt. Eng. Geol. **218**, 77–89 (2017b)

Khalil, M.M., Tokunaga, T., Yousef, A.F.: Insights from stable isotopes and hydrochemistry to the Quaternary groundwater system, south of the Ismailia canal, Egypt. J. Hydrol. **527**, 555–564 (2015)

Lokmer, I., Herak, M., Panza, G.F., Vaccari, F.: Amplification of strong ground motion in the city of Zagreb, Croatia, estimated by computation of synthetic seismograms. Soil Dyn. Earthq. Eng. **22**(2), 105–113 (2002)

Magrin, A., Gusev, A.A., Romanelli, F., Vaccari, F., Panza, G.F.: Broadband NDSHA computations and earthquake ground motion observations for the Italian territory. Int. J. Earthq. Impact Eng. **1**(1–2), 131–158 (2016)

Maklad, M., Yokoi, T., Hayashida, T., ElGabry, M.N., Hassan, H.M., Hussein, H.M., Fattah, T.A.: Site characterization in Ismailia city, Egypt using seismic ambient noise array. J. Eng. Geol. (2020)

Mourabit, T., Elenean, K.A., Ayadi, A., Benouar, D., Suleman, A.B., Bezzeghoud, M., Cheddadi, A., Chourak, M., ElGabry, M.N., Harbi, A., Hfaiedh, M.: Neo-deterministic seismic hazard assessment in North Africa. J. Seismol. **18**(2), 301–318 (2014). https://doi.org/10.1007/s10950-013-9375-2

Molchan, G., Kronrod, T., Panza F.: Hot/cold spots in Italian macroseismic data. Pure Appl. Geophys. **168**, 739–752 (2011). https://doi.org/10.1007/s00024-010-0111-3

Panza, G.F., Romanelli, F., Vaccari, F.: Seismic wave propagation in laterally heterogeneous anelastic media: theory and applications to seismic zonation. Adv. Geophys. **43**, 1–95 (2001)

Panza, G.F., La Mura, C., Peresan, A., Romanelli, F., Vaccari, F.: Chapter three-seismic hazard scenarios as preventive tools for a disaster resilient society. Adv. Geophys. **53**, 93–165 (2012)

Said, R.: The Geology of Egypt. A.A. Balkema, Rotterdam (1990)

Trifunac, M.D.: Earthquake response spectra for performance based design-a critical review. Soil Dyn. Earthq. Eng. **37**, 73–83 (2012). https://doi.org/10.1016/j.soildyn.2012.01.019

Zuccolo, E., Vaccari, F., Peresan, A., Panza, G.F.: Neo-deterministic and probabilistic seismic hazard assessments: a comparison over the Italian territory. Pure. Appl. Geophys. **168**(1–2), 69–83 (2011)

Relationships of the Physical and Mechanical Properties Obtained in Ignimbrites from a Tunnel in Morelia, Mexico

Carlos Luz-Martínez, Eleazar Arreygue-Rocha, and Luisa Equihua-Anguiano(✉)

UMSNH University, Ciudad Universitaria, Gral. Francisco J. Mugica S/N, 58030 Morelia, Michoacan, Mexico
jc.luz.mtz@gmail.com, elearreygue@gmail.com, nicteea@yahoo.com.mx

Abstract. Tunnel construction is important to reduce the length of the geometric road design as well as the environmental impact. However, its construction is one of the most complex challenges due to the variability of the geotechnical properties. Stability of the tunnels in rock can be compromised for the rock mass deterioration which can cause faults during and after construction. Deterioration rate depends on the rock properties together with the environmental factors. Rock deterioration predictions can be achieved by *in situ* explorations and by performing laboratory tests. One important test is the slake durability, since it considers wetting and drying cycles subjected to the material wastage. This paper is focused on determining and obtaining relationships of the durability of ignimbrites from Morelia, Mexico. The ignimbrite was obtained from one geotechnical borehole of 100 m of length on the axis of a road tunnel and the physical and mechanical properties of the ignimbrites were determined, hence the deterioration of the rocks. This last, was measured using five cycles instead of the two cycles considered in the Normative. Results obtained show that from the third cycle, a constant deterioration of the rock is observed impacting in the slake durability index (I_d) and furthermore, it depends in the rock hardness. The I_d after the fifth cycle (I_{d5}) was associated to estimate the physical and mechanical properties of the ignimbrites. It was found that the increase in I_{d5} correlates with the increase in density, compressive strength, tensile strength, P-wave velocity and decrease in porosity and absorption.

1 Introduction

Rock mass instability is associated with the rock deterioration (Brox 2017), nevertheless in tunnels design it has been given low importance, because it is in general based on failure mechanisms. Rock deterioration is produced by the perturbation in the moment of the tunnel excavation since the confinement release causes a stress redistribution (Gerber and Scheidegger 1969). Furthermore, the rock is exposed by the excavation to the environmental conditions such as temperature and humidity (Feld 1966), changing the original rocks characteristics. In a rock mass, the correct determination of the resistance and deformability, considering the rock as the combination of intact rock and discontinuities (Narimani-Dehnavi and Sadeghi 2017)) through *in situ* and laboratory tests is

H. Shehata and M. Badr (Eds.): *Advancements in Geotechnical Engineering*, SUCI, pp. 79–90, 2021.
https://doi.org/10.1007/978-3-030-62908-3_6

difficult, thereby mechanical properties are preferably determined from the matrix rock, using empirical correlations and models based on the definition of a geomechanical quality. One index that takes in account the quality of the rock in a qualitative form (Tatiya 2005) is the Rock Quality Index (RQD) and it was proposed by Deere and Deere (1988). In the other hand, it is necessary to obtain the weathering and alteration of the rocks, since those have an important influence on the mechanical properties of the material (Boyd, 1975). Other important parameter in tunnels design is the rock mass rating (RMR) classification that has been developed by Bieniawski (1989) and is based on field measurements of more than 300 cases of tunnels and foundations worldwide. This classification takes into account the characteristics of the discontinuities, the resistance to uniaxial compression and the RQD. In the same way, the Q classification or the NGI quality index (Barton et al. 1974), that is based on the analysis of the 200 tunnel cases from which a correlation is established between the quantity and type of permanent support and the quality of the rock mass Q, with respect to the stability of the tunnel, takes into account the degree of fracture of the rock mass, the characteristics of the discontinuities, the presence of water and the active stress.

Deterioration begins to have importance when problems of stability in tunnels are presented as those described by Lamas (1993), that describes several accidents and other damages in hydraulic tunnels. Table 1 shows different kinds of damages that correspond to typical situations that occur in civil works. Most of the problems occurred in tunnels with concrete lining or without any type of lining. Figure 1 shows distribution of the deterioration cases expressed in percentage terms, for the situations of hydraulic tunnels without lining and with concrete lining and according to Table 1. Furthermore, Harrison and Hudson (2000) point out that there are factors that influence the stability of the rock mass, especially the deterioration of the rock, due to exposure and other time-dependent effects. Although weathering processes can be slow, they depend on the type of rock and the situation. Water introduces new chemicals that induce the decomposition of the minerals in the rock make-up, but also acts to transport the decomposed and disintegrated waste from the mother rock (Hencher 2015).

Considering the previous statements, this article presents a study to obtain the deterioration in ignimbritic rock cores of a constructed tunnel in Morelia, Mexico. For that, the properties were determined by using different laboratory standardized tests, such as: density, absorption, porosity, uniaxial compressive strength, tensile strength, P-wave velocity and durability. This last was developed with a modification respect to the international normative and that allowed to obtain specific results in the local ignimbrites as well as, correlations among the different properties obtained in the laboratory.

Table 1. Classes of classification of deterioration (Lamas 1993)

Class	Description of deterioration
A	Inadequate confinement, leading to excessive leakage, hydraulic jacking or rock mass instability, including landslides or uplift
B	Specific geological features of high hydraulic conductivity, leading to leakage, hydraulic jacking or rock mass instability, including landslides or uplift
C	Deterioration of rock mass, namely due to erosion of seams, dissolution and swelling, leading to excessive leakage, rock falls or rock mass instability
D	Excessive water pressure against impermeable barriers, such as seams or clay filled faults, leading to rock mass movement and instability, including landslides
E	Deformable rock mass, inefficient grouting or deficient construction, leading to failure of the lining, namely owed due to the internal water pressure
F	Buckling of steel linings caused by external pressure of water or grouting
G	Dynamic fluctuations of water pressure

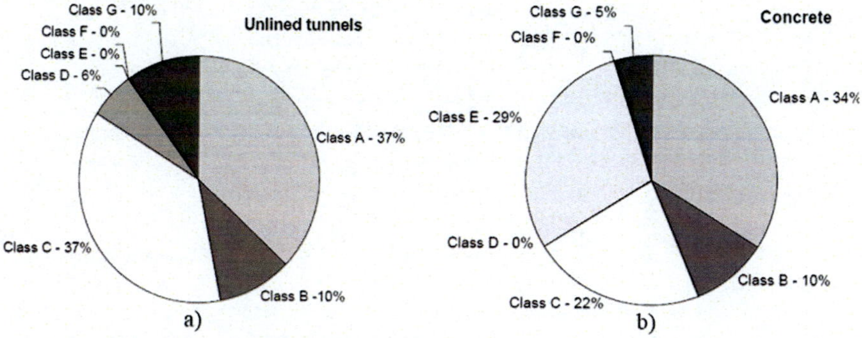

Fig. 1. Distribution of the classes of deterioration for a) unlined and b) concrete pressure tunnels (Ribeiro and Sousa, 2006).

2 Case Study

Road tunnel studied is located in the city of Morelia, Mexico. Two boreholes of 100 m of depth were performed with a diameter of 47.6 mm (NQ coring). In Fig. 2 the scheme of the two drilling locations is shown. Nevertheless, in this paper only the results of the "Core Drilling 2" are presented, where the tunnel is among 70 and 90 m of depth. From previous studies realized in this zone (Arreygue-Rocha et al, 2002), it was possible to correlate with a better precision, the physical ignimbrite characteristics with the recent sampling achieved. Figure 3 presents the stratigraphic sequence characterized by Arreygue-Rocha et al. (2002). In this Figure it is appreciated that the formation is an ignimbrite rock and one hypothesis of its origin, is the material deposition caused by intense and frequents air fall or flows of pyroclastic material and small clouds of intensely hot volcanic material that give the origin name of the ignimbrite rocks (Marshall, 1935).

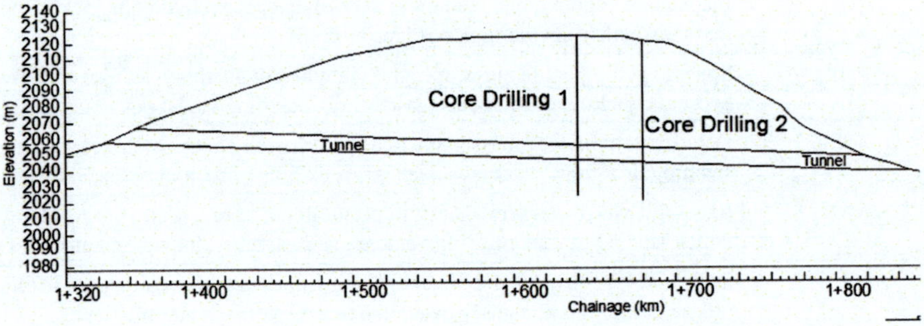

Fig. 2. Scheme of the location of the tunnel rock boreholes.

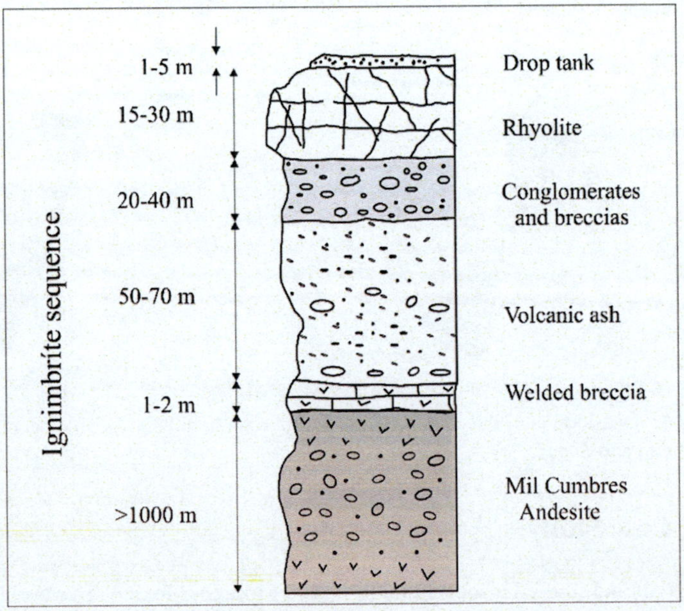

Fig. 3. Stratigraphic sequence of the Morelia, City Mexico (Arreygue-Rocha et al., 2002).

3 Sampling and Methods

3.1 Laboratory Tests

Samples used were selected and prepared in accordance with the ASTM D4543-08e1 (2017). Table 2 shows the laboratory tests that were carried out on the ignimbrite rock cores, as well as the equations to obtain the different physical and mechanical property values as density, absorption, porosity, point load index, uniaxial compression strength, ultrasonic pulse wave velocity, tensile strength and durability index.

Table 2. Laboratory tests on the ignimbrite rock cores

Eq.	Test	Symbol	Expression	Where
1	Density	ρ *(g/cm3)*	*m/v*	*m = sample mass (g)* *v = sample volume (cm3)*
2	Absorption	*Absorption (%)*	*(B − A)/A*	*B = mass of saturated sample (g)* *A = mass oven-dried sample (g)*
3	Porosity	*n* *(-)*	$(V_v/V)\cdot 100$	V_v *= voids volume* *V = total volume*
4	Point Load Index	I_s *(MPa)*	P/D_e^2	*P = failure load (N)* D_e *= equivalent core diameter (mm)*
5	Uniaxial compression strength	σ_c *(MPa)*	$f\cdot I_s$ $f = (D_e/50)^{0.45}$	*f = correction factor* I_s *= point load index (MPa)*
6	Ultrasonic pulse wave velocities	V_P *(m/s)*	L_P/T_P	L_P *= pulse travel distance (m)* T_P *= effective pulse-time travel (s)*
7	Tensile strenght	σ_t *(MPa)*	$(2\cdot P)/(\pi\cdot t\cdot D)$	*P = maximal applied load (MPa)* *t = specimen thickness (mm)* *D = specimen diameter (mm)*
8	Durability index	I_d *(%)*	$\frac{(W_F - C)}{(B - C)} 100$	W_F *= mass of the drum + sample oven-dried after cycle (g)* *B = mass of the drum + sample oven-dried before cycle (g)* *C = drum mass (g)*

3.2 Geomechanical Clasifications

Geomechanical classifications were performed in all the rock units in which, the "Core Drilling 2" was separated as is presented in Fig. 4. The rock mass rating (RMR) was obtained using the Eq. (9), the rock mass quality (Q) was obtained using the Eq. (10) and the rock quality (RQD) was obtained with the Eq. (11).

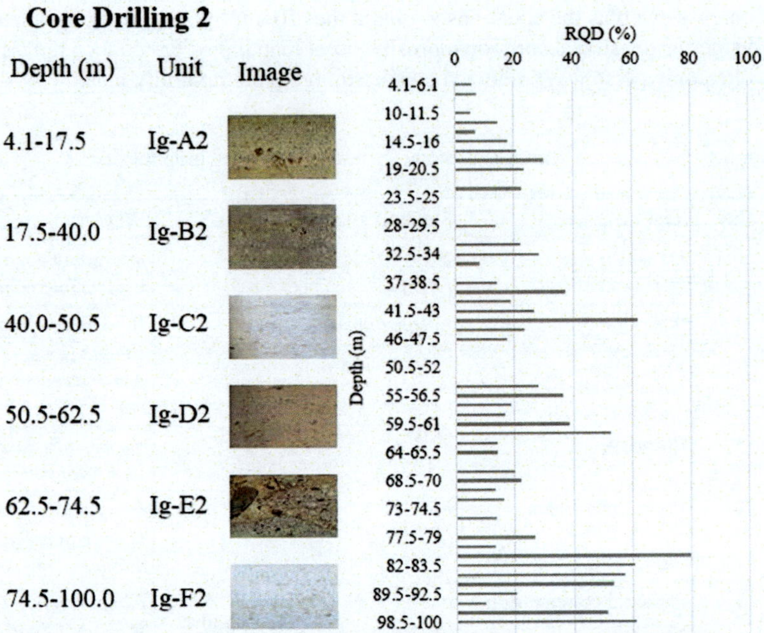

Fig. 4. Unit division and RQD variation for the "Core Drilling 2" sampling

$$RMR = R_c + R_{RQD} + R_d + R_s + R_u \tag{9}$$

where: R_c is the unaltered rock resistance, R_{RQD} is the rock quality designation (RQD), R_d is the spacing of discontinuities, R_s is the condition of discontinuities (length, persistence, separation, smoothness, infilling, alteration/weathering) and R_u are the groundwater conditions.

$$Q = \frac{RQD}{J_n} \cdot \frac{J_r}{J_a} \cdot \frac{J_w}{SRF} \tag{10}$$

where: RQD is the rock quality designation, J_n is the score of the joint sets number, J_r is the score for the roughness of joint surface, J_a is the score for the clay filling or alteration degree, J_w is the groundwater pressure effects scores and the stress reduction factor and SRF is the score for faulting, strength and stress ratios in hard rocks and squeezing or swelling.

$$RQD = \frac{\sum length\ of\ core\ pieces\ >\ 10\,\text{cm}}{total\ lenght} \cdot 100 \tag{11}$$

where: $\Sigma =$ Summation of the length of core pieces that are >100 mm (4 in.) measured along the centerline and the total core length.

4 Results

4.1 Recognition "Core Drilling 2"

Based on the core evaluation conjointly with the laboratory tests, the borehole was divided in six units (see Fig. 4), which were named Ig-A2, Ig-B2, Ig-C2, Ig-D2, Ig-E2 and Ig-F2 respectively. Table 3 shows the material description and its main physical characteristics. The RQD is presented in Fig. 4, results show that the rock quality goes from the bad to very bad quality. Values obtained are less than RQD $< 50\%$ which indicates a high fracturing degree.

Table 3. Survey recognition of the "Borehole 2" and nomenclature of the classification.

Unity	Name	Description
1	Ig-A2	Ignimbrite rocks with light brown color, with presence of large pumice particles with a considerable degree of fracturing and with a highly weathered degree in the surface
2	Ig-B2	Ignimbrite changes to a dark-gray color, pumice are still present but in smaller quantities and sizes. Degree of welding increases and the rock has an almost massive appearance
3	Ig-C2	Similar characteristics to the Ig-B2, with a change in the light gray-color. Presence of pumice is more visible and the weathering is only observed in discontinuities or areas with intense fracturing
4	Ig-D2	Very solded ignimbrite with pink color. It has a massive appearance nevertheless, it is fractured. Pumice lenses are observed and not presents weathering except in discontinuities
5	Ig-E2	There is a very noticeable material change. Clasts of different sizes are observed surrounded by a sandy matrix with different cementation degrees
6	Ig-F2	Is a poorly welded material with the presence of large pores and the fracturing degrees diminish. Texture is markedly porphyritic; it can be seen that the bonding of the particles is stronger

4.2 Density (ρ), Absorption, Porosity (N) and Uniaxial Compression Strength (σ_C)

Figure 5 shows the properties variation with depth of a) density (ρ), b) absorption, c) porosity (n) and the uniaxial compression strength (σ_c). It can be seen that in the tunnel depth location these properties diminish for ρ, and the σ_c cases, oppositely to the n and the absorption values. From the surface to 70 m borehole depth, this trend variation is conservatively constant except in the units denoted as Ig-D2 and Ig-E2, where these properties present an erratic variation some times, but it is possible to define a constant variation with depth.

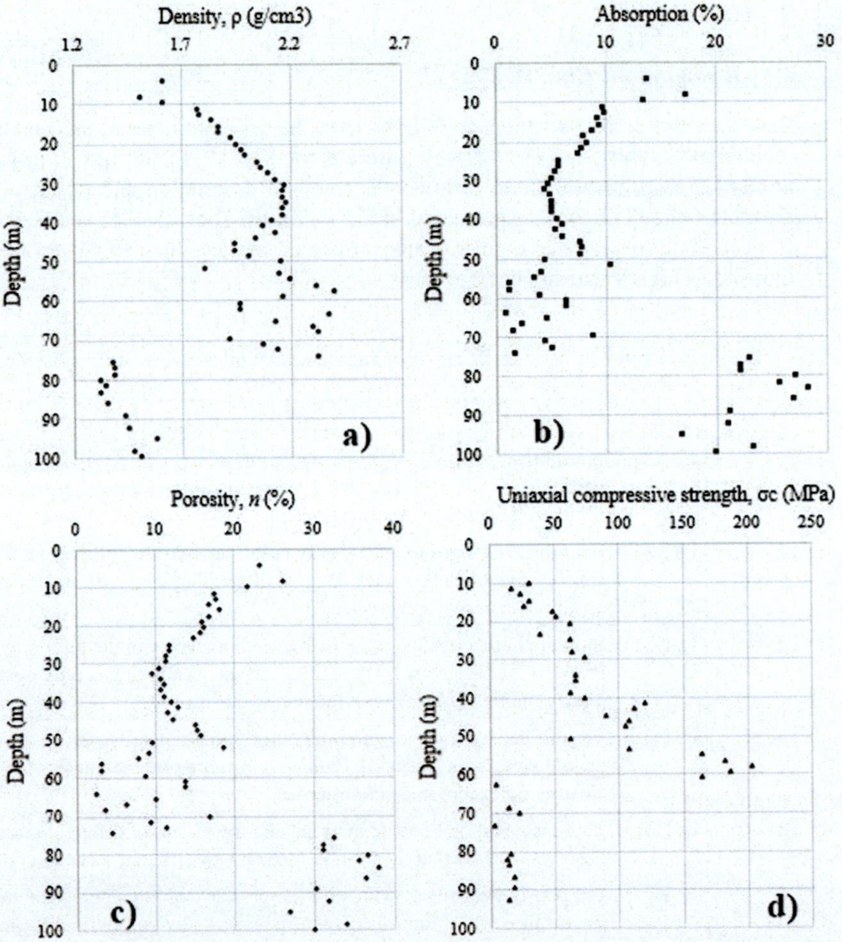

Fig. 5. "Core Drilling 2" Depth variation of the a) Density ρ, b) Absorption, c) Porosity (n) and uniaxial compression strength (σ_c) properties.

Figure 6 shows the relationships among ρ, n and Absorption as well as, the expressions of the trend line, which follows the same tendency. It can be seen that there is a relationship between these three properties. When density values increases, absorption and porosity decreases, which means that ρ is proportional to the n and Absorption. Ignimbrites with high densities have low Absorption and porosity percentages and for this type of rocks this characteristic these relationships are met all time.

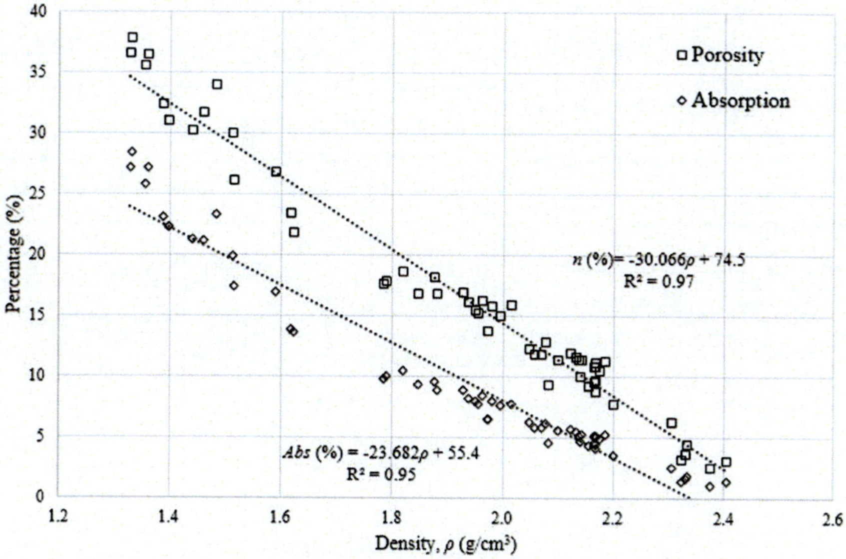

Fig. 6. Density ρ vs Porosity n relationships in the "Core Drilling 2".

In Fig. 7, it can be see the relationship trend among the ρ vs σ_c. In this case a greater dispersion is observed, however an expression is presented. A point that can be discussed is that for the case of high resistance σ_c values, a low degree of weathering is presented, nevertheless not for the high density rock values observed. Hence, it is necessary to find other parameters that exhibit a better relation among the alteration state or with the susceptibility to deterioration of the ignimbritic rocks. Figure 8 exposes the V_p and the ρ relation, in this last case the trend follows practically a lineal tendency. Highest P wave velocities correspond to the ignimbrite samples with the highest densities, confirming the strong influence of density on the P wave velocity. Unlike the relationships between uniaxial compressive strength with the density and the porosity, the P wave velocity has a better correlation with these properties.

4.3 Ignimbrites Durability (Id_5)

Figure 9 shows the ignimbrite durability, obtained performing five cycles of wastage (I_{d5}) on the six units studied. It is observed that the units that presented a loss of material greater than 10% (Ig-E and Ig-F) in the second cycle, have more unfavorable behavior against deterioration in subsequent cycles. This would be an indicator in this type of rock, of when it is necessary to perform additional cycles to the second cycle stipulated in the normative. In general samples with losses less than 10% in the second cycle retain an I_d greater than 90% in the fifth cycle, except for the Ig-E2 unit with an I_d of 89%. These results show the importance of the weathering rock degree in the durability index and the possibility to add more cycles to the test. Table 4 shows the I_d obtained with Eq. (8) in each cycle realized.

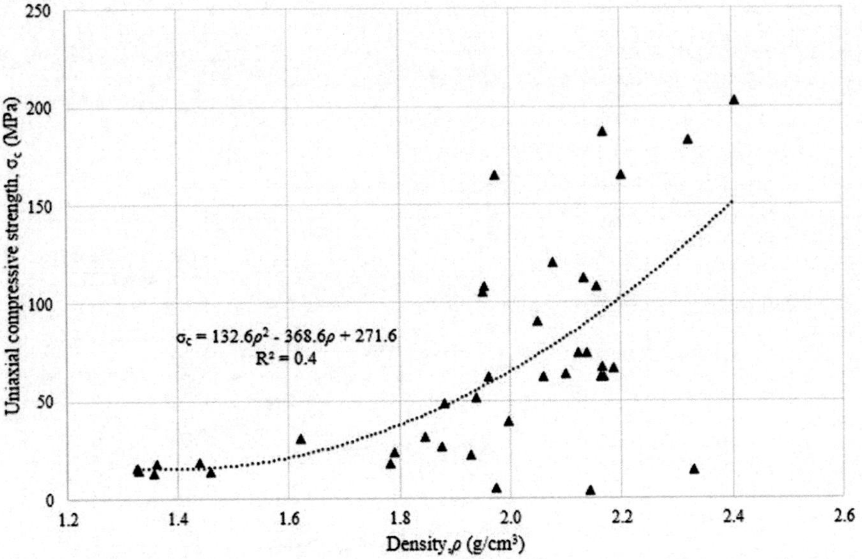

Fig. 7. Density ρ vs uniaxial compression strength σ_c relationship in the "Core Drilling 2".

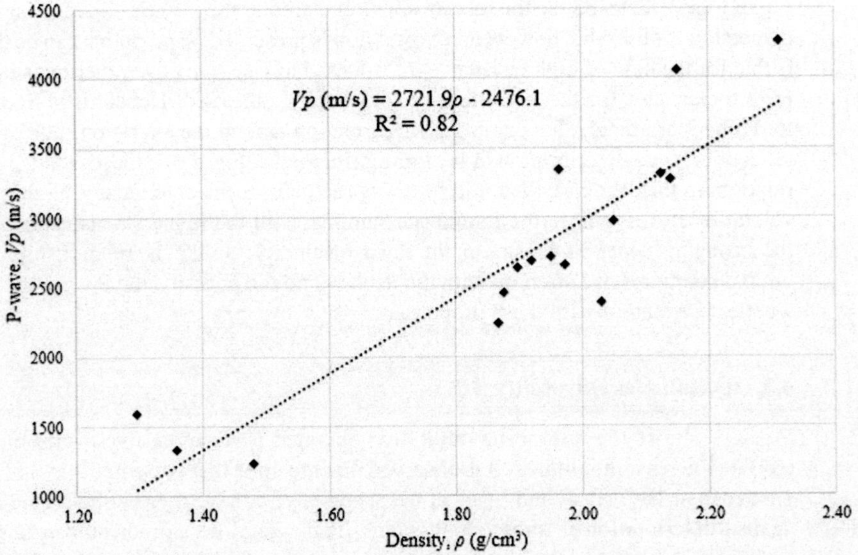

Fig. 8. Ultrasonic pulse wave velocities Vp vs Density ρ relationships in the "Core Drilling 2".

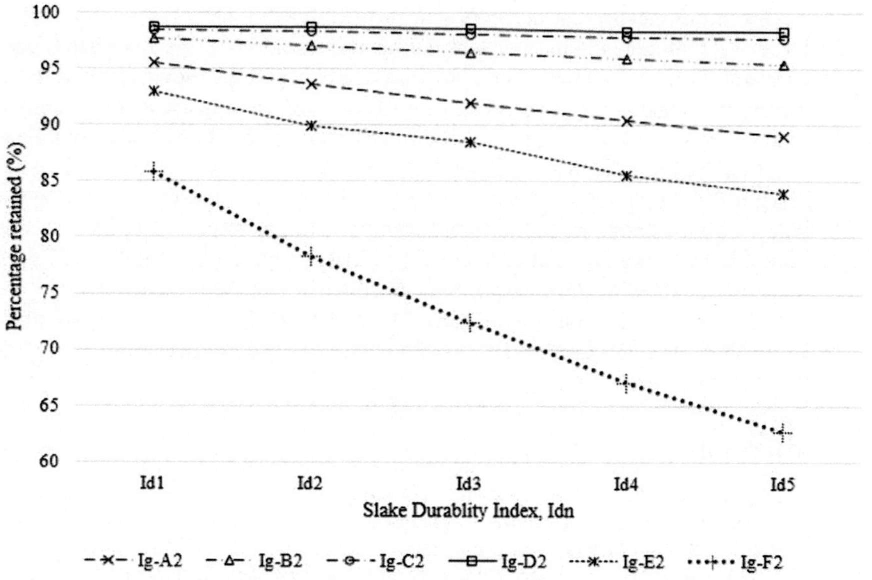

Fig. 9. Ignimbrites durability performing 5 cycles of wastage in the "Core Drilling 2".

Table 4. Durability index Id, in the six units of the "Core Drilling 2" for the five cycles.

Unit	Id_1	Id_2	Id_3	Id_4	Id_5
Ig-A2	95.5	93.6	92.0	90.4	89.0
Ig-B2	97.7	97.1	96.4	95.9	95.4
Ig-C2	98.5	98.4	98.0	97.8	97.6
Ig-D2	98.7	98.7	98.6	98.3	98.3
Ig-E2	92.9	89.9	88.4	85.5	83.9
Ig-F2	85.2	78.2	72.4	67.0	62.7

5 Conclusions

This paper is focused in the study to the weathering process and its influence on the durability of an ignimbritic sequence. Physical and mechanical properties to characterized ignimbrites that compose the stratigraphy of a road tunnel constructed in Morelia, Mexico are presented. The principal contribution of this work is the obtainment of the ignimbrite durability index I_d, through the application of five cycles instead of the two cycles fixed in the standardized test. Results showed that the weathering degree in the ignimbrites have a big influence in results obtained of the I_d. These suggest that depending of the rock state, it can be necessary to carrying out more cycles to avoid future failures in the constructions.

From the mechanical and physical properties were obtained relationships among these were obtained. This permits the characterization of the material founded and compares with studies developed previously. Relationships showed a good match trend among the density, the porosity and the absorption. In the case of the density and the uniaxial compression strength a greater dispersion in results was observed, hence, it is necessary to find other parameters that exhibit a better relation among the alteration state or with the susceptibility to deterioration of the ignimbritic rocks. In the case of the P wave velocities and the density relationship, the trend follows practically a lineal tendency. Highest P wave velocities correspond to the ignimbrite samples with the highest densities, confirming the strong influence of density on the P wave velocity.

This characterization is important, due to constructions that can be take place in the future. Results were compared with other works realized in other countries with presence of ignimbrite rocks.

References

Arreygue-Rocha, E., Garduño-Monroy, V.H., Canuti, P., Casaglie, N., Iotti, A., Sergio, C.: Análisis geomecánico de la inestabilidad del escarpe La Paloma, en la Ciudad de Morelia, Michoacán. México. Revista Mexicana de Ciencias GeolAóngáicliassi **19**(2), 91–106 (2002)

ASTM D4543-08e1: Preparing Rock Core as Cylindrical Test Specimens and Verifying Conformance to Dimensional and Shape Tolerances. ASTM International, West Conshohocken (2017)

Barton, N., Lien, R., Lunde, J.: Engineering classification of rock masses for the design of tunnel support. Rock Mech. **6**(4), 189–236 (1974)

Bieniawski, Z.: Engineering Rock Mass Classifications. John Wiley & Sons, Pennsylvania (1989)

Boyd J. M.: The Interpretation of Geological Structure for Engineering Design in Rock. University of London, PhD thesis (1975)

Brox, D.: Practical Guide to Rock Tunneling. CRC Press, London (2017)

Deere D.U., Deere D.W.: The rock quality designation (RQD) index in practice. In: Rock Classification Systems for Engineering Purposes, pp. 91–101 (1988)

Feld, J.: Rock movements from load release in excavated cuts. In: Proceedings 1st International Congress Society Rock Mechanics, Lisbon, pp. 139–140 (1966)

Gerber E., Scheidegger A.: Stress-induced weathering of rock masses. Eclogae Geol. Helv. **62**, 401–406 (1969)

Harrison, J.P., Hudson, J.A.: Engineering Rock Mechanics: Part 2. Pergamon, The Netherlands (2000)

Hencher, S.: Practical Rock Mechanics. CRC Press, Boca Raton, FL (2015)

Jayalekshmi, B.R., Chinmayi, H.K.: Effect of soil stiffness on seismic response of reinforced concrete buildings with shear walls. Innovative Infrastruct. Solutions **1**(1), 1–18 (2016). https://doi.org/10.1007/s41062-016-0004-0

Lamas, L.N.: Contributions to understanding the hydromechanical behaviour of pressure tunnels. PhD thesis, London: Imperial College (1993)

Marshall, P.: Acid rocks of the taupo-rotorua volcanic district. Roy. Soc. NZ **64**(3), 323–366 (1935)

Narimani-Dehnavi, R., Sadeghi, M.: Deterioration of weak rocks over time and its effect on designing tunnel support systems. Eng Geol Environ. **78**, 1045–1056 (2017). https://doi.org/10.1007/s10064-017-1154-9

Ribeiro e Sousa, L.: Learning With Accidents And Damage Associated To Underground Works. Taylor & Francis Group, London (2006)

Tatiya, R.: Civil Excavations And Tunnelling – A Practical Guide. Thomas Telford, London, Great Britain (2005)

Implementation of a Single Hardening Constitutive Model for 3D Analysis of Earth Dams

Juan Huang and Celso Romanel(✉)

Department of Civil and Environmental Engineering, Pontifical Catholic University of Rio de Janeiro, Rio de Janeiro, Brazil
jandreshuang@outlook.es, romanel@puc-rio.br

Abstract. In the practice of geotechnical engineering, reliable information about soil properties are often available in terms of strength parameters but data on the deformability characteristics of the materials are known to a lesser extent. For situations where the analysis of soil resistance is a condition of the project, the traditional Mohr-Coulomb model is usually applied, but there are several other situations where the anticipated estimate of the level of deformations induced by the project is an essential engineering information. In this research the single hardening model proposed by Lade and Kim was implemented in the FLAC 3D finite difference program in order to investigate the three-dimensional geomechanical behavior of earth dams during the construction stage. Numerical analyses were carried out considering 2D and 3D models to study the influence of the intermediate stress on the geomechanical behavior of earth dams. The computed results were also compared with those obtained using the traditional Mohr-Coulomb model to put in evidence the advantages using the single hardening constitutive model when analyzing geotechnical problems.

1 Introduction

Numerical simulations of earth dams are influenced by the soil parameters obtained from field or laboratory tests as well as by the constitutive model chosen to represent the mechanical behavior of the soil mass. Traditional constitutive relationships, such as the well-known Mohr-Coulomb soil model, present several difficulties when simulating the elastoplastic behavior of soils before failure. During the last 4 decades several advanced soil models were proposed in the literature as attempts to improve predictions of soil deformation, which is a key factor in many engineering projects involving excavation, foundation settlement, consolidation, retaining structures etc. One of such advanced models was proposed by Kim and Lade (1988) and Lade and Kim (1988a) considering a single hardening elastoplastic surface that, according to the authors, could be used to simulate the stress-strain behavior of all frictional materials (soils, rocks, concrete) under any combination of loadings/unloading in drained or undrained analysis. In the present research the implementation of Lade & Kim soil model was made in the 3D FLAC finite difference commercial software based on an explicit integration scheme developed by Jakobsen and Lade (2002).

H. Shehata and M. Badr (Eds.): *Advancements in Geotechnical Engineering*, SUCI, pp. 91–105, 2021.
https://doi.org/10.1007/978-3-030-62908-3_7

2 Single Hardening Constitutive Model

Elastic Behavior

The elastic strains are calculated by Hooke's generalized law considering a constant Poisson's ratio ($0 \leq v \leq 0.5$). A formulation based on the energy conservation principle was developed by Lade and Nelson (1987) to express the variation of the modulus of elasticity with the stress state, but independent of the stress path (hyperelastic formulation).

$$E = Mp_a\left[\left(\frac{I_1^2}{p_a}\right) + 6\left(\frac{1+\nu}{1-2\nu}\right)\frac{J_2^{'}}{p_a}\right]^{\lambda} \tag{1}$$

where the modulus number M and the exponent λ are dimensionless constants, I_1 is the first Cauchy invariant of the stress tensor and J'_2 is the second invariant of the deviatoric stress tensor and p_a represents the atmospheric pressure.

Lade and Kim (1988b) also indicated the unloading-loading elastic modulus E_{ur} proposed in the hyperbolic material model (Duncan and Chan 1970) to calculate the elastic strain increments:

$$E_{ur} = K_{ur}p_a\left(\frac{\sigma_3}{p_a}\right)^n \tag{2}$$

where K_{ur} and n are dimensionless constants, σ_3 is the actual confining stress.

Failure Criterion

The failure criterion is expressed in terms of the Cauchy stress invariants I_1 and I_3 as defined below:

$$f_n = \left(\frac{I_1^3}{I_3} - 27\right)\left(\frac{I_1}{p_a}\right)^m \tag{3a}$$

$$f_n = \eta_1 \text{ at failure} \tag{3b}$$

where η_1 and $m \leq 1.979$ are dimensionless constants.

In the principal stress space, the failure criterion appears has an asymmetric bullet with apex at the origin and cross-sectional shape (octahedral plane) triangular with smoothly rounded vertices.

Plastic Potential and Flow Rule

The plastic potential function g_p is mathematically written as function of the three Cauchy stress invariants:

$$g_p = \left(\psi_1\frac{I_1^3}{I_3} - \frac{I_1^2}{I_2} + \psi_2\right)\left(\frac{I_1}{p_a}\right)^{\mu} \tag{4}$$

where the parameters ψ_1, ψ_2 and μ are dimensionless constant numbers. The parameter ψ_1 is a weighting factor between the triangular and circular shape of the plastic function in the octahedral plane, ψ_2 controls where this function intercepts the hydrostatic axis and μ models the curvature of the meridians. Value of the material parameter ψ_1 can be determined through the following empirical correlation with the curvature parameter m of Eq. 3a:

$$\psi_1 = 0.00155\, m^{-1.27} \tag{5}$$

In the Lade and Kim model the plastic strain increment $d\varepsilon^p$ is calculated from the nonassociated flow rule

$$d\varepsilon^p = d\lambda_p \frac{\partial g_p}{\partial \sigma} \tag{6}$$

where $d\lambda_p$ is a positive constant of proportionality (plastic multiplier).

Yield Criterion

The isotropic yield surfaces, derived from surfaces of constant plastic work (Lade and Kim 1988a), are expressed as:

$$f_p = f_p'(\sigma) - f_p''(W_p) \leq 0 \tag{7a}$$

with

$$f_p' = \left(\psi_1 \frac{I_1^3}{I_3} - \frac{I_1^2}{I_2}\right)\left(\frac{I_1}{p_a}\right)^h e^q \tag{7b}$$

where h is a material constant, determined considering two points (one on the hydrostatic axis, another on the failure surface) on the basis that the plastic work that the plastic surface W_p is constant along a yield surface. On the other hand, the parameter q varies with the stress level S from 0 (hydrostatic axis) to unity (failure surface) according to:

$$q = \frac{\alpha S}{1 - (1 - \alpha)S} \tag{7c}$$

admitting α as a constant number.

The hardening function $f_p''(W_p)$ can be written as:

$$f_p'' = \left(\frac{1}{D}\right)^{1/\rho}\left(\frac{W_p}{p_a}\right)^{1/\rho} \tag{7d}$$

the values of ρ and D are constants given by:

$$\rho = p/h \tag{7e}$$

$$D = \frac{C}{(27\psi_1 + 3)^\rho} \tag{7f}$$

where C and p are parameters from modeling the plastic work through isotropic compression:

$$W_p = Cp_a\left[\frac{I_1}{p_a}\right]^p \tag{7g}$$

which can be generalized as:

$$W_p = p_a D\left(f_p''\right)^{\rho} \tag{7h}$$

Considering that the plastic potential function is a homogeneous function of the μth degree (Wylie and Barret 1982) then the scalar factor of proportionality $d\lambda_p$ (Eq. 6) can be related to the plastic work increment dW_p and the plastic potential function g_p according to:

$$d\lambda_p = \frac{dW_p}{\mu g_p} \tag{8a}$$

where dW_p is obtained by differentiation of Eq. 7h.

$$dW_p = p_a D\rho\left(f_p''\right)^{\rho-1} df_p'' \tag{8b}$$

3 Numerical Implementation

Incremental Formulation

The numerical implementation of the elastoplastic model into the finite difference program FLAC3D was done through generation of a dynamic link library (DLL) considering the constitutive equations written in their incremental form by Jakobsen and Lade (2002).

Elastoplastic stiffness matrix C^{ep}

$$d\sigma = C^{ep}d\varepsilon = \left(C^e - C^p\right)d\varepsilon \tag{9a}$$

where the elastic stiffness matrix C^e

$$C^e = \frac{E}{(1+\nu)(1-2\nu)}\begin{bmatrix} 1-\nu & \nu & \nu & 0 & 0 & 0 \\ \nu & 1-\nu & \nu & 0 & 0 & 0 \\ \nu & \nu & 1-\nu & 0 & 0 & 0 \\ 0 & 0 & 0 & 1-2\nu & 0 & 0 \\ 0 & 0 & 0 & 0 & 1-2\nu & 0 \\ 0 & 0 & 0 & 0 & 0 & 1-2\nu \end{bmatrix} \tag{9b}$$

and the plastic stiffness C^p is defined in terms of the derivatives of the plastic potential g_p, the derivatives of the yield function f_p and the hardening modulus H.

$$C^p = \left[\frac{C^e \left(\frac{\partial g_p}{\partial \sigma}\right) \left(\frac{\partial f_p}{\partial \sigma}\right)^T C^e}{\left(\frac{\partial f_p}{\partial \sigma}\right)^T C^e \left(\frac{\partial g_p}{\partial \sigma}\right) + H} \right] \tag{9c}$$

$$H = \frac{\partial f_p''}{\partial W_p} \sigma^T \frac{\partial g_p}{\partial \sigma} = \frac{\partial f_p''}{\partial W_p} \mu g_p \tag{9d}$$

Expressions of the derivatives implied in Eqs. 9c and 9d can be found in the appendix.

An algorithm is needed to obtain the updated state of stress and hardening parameter due to a strain increment. The forward Euler scheme may be used for elastoplastic stress-strain calculations but it has some limitations regarding numerical accuracy. FLAC3D software, because of its explicit, dynamic solution scheme relies on this type of integration scheme but, in this research, it has been used a modified forward Euler scheme with error control, developed by Sloan (1987) as indicated in Fig. 1, in order to enhance the accuracy of the explicit algorithm.

Validation 1 - Sacramento River Sand

Numerical simulations of triaxial compression tests were performed to validate the implemented elastoplastic constitutive model into FLAC3D considering typical geotechnical parameters for loose Sacramento river sand reported by Lade (2005) in Table 1. The computed results are shown in Fig. 2, where it can be observed strong agreement with the experimental data for that 4 levels of confining stresses ($\sigma_3 = 94$ kPa, 200 kPa, 450 kPa, 1265 kPa).

Validation 2 - Roadford Dam

The Roadford dam is a 41 m high low-grade rockfill with an asphaltic concrete membrane on its upstream face to provide waterproofing protection. An important design characteristic of the dam is the junction of the asphaltic membrane and the cutoff structure at the upstream toe of the dam. To reduce differential movements in this region a zone of stiffer sand waste fill was used immediately against the cut-off structure. Drainage measures included drainage layers located on top of the foundation and the upstream face of the dam just beneath the asphaltic concrete membrane. Details of the design and the materials used in the construction were given by Wilson and Evans (1990). Large scale tests were carried out at the Building Research Establishment (BRE) to determine the relevant properties of the low grade rockfill and sand waste fill. These included investigations of compressibility in a 1 m diameter oedometer and determination of rockfill strength/compressibility by drained triaxial compression tests on 0.23 m diameter samples. The stiffness and strength parameters derived from these tests were used by Kovacevic (1994) in finite element analyses to predict the behavior of the dam, both during construction and reservoir impounding, considering the single hardening constitutive model with parameter values shown in Table 2. Bulk unit weights were $\gamma = 22.75$ kN/m^3 for the rockfill and $\gamma = 18.35$ kN/m^3 for the sand waste fill. The foundation was admitted as a linear-elastic material with values of Young's modulus $E = 225$ MPa and Poisson's ratio $\nu = 0.20$.

$$Initial\ state: \sigma_0, W_{p,0}, \Delta\varepsilon, \delta\varepsilon_0 = \Delta\varepsilon, q = 1$$
$$Strain\ increment\ i = 1, 2, \ldots, n$$
$$\delta\varepsilon_i = \min\left[q\delta\varepsilon_{i-1}, 1 - \sum_{j=1}^{i-1} \delta\varepsilon_j\right]$$
$$do$$
$$\delta\sigma_i^I = C^{ep}(\sigma_{i-1}, W_{p,i-1})\delta\varepsilon_i$$
$$\delta W_{p,i}^I = \delta\lambda_p(\sigma_{i-1}, W_{p,i-1}, \delta\varepsilon_i)\sigma_{i-1}\frac{\partial g}{\partial\sigma}$$
$$\delta\sigma_i^{II} = C^{ep}(\sigma_{i-1} + \delta\sigma_i^I, W_{p,i-1} + \delta W_i^I)\delta\varepsilon_i$$
$$\delta W_{p,i}^{II} = \delta\lambda_p(\sigma_{i-1} + \delta\sigma_i^I, W_{p,i-1} + \delta W_i^I, \delta\varepsilon_i)(\sigma_{i-1} + \delta\sigma_i^I)\frac{\partial g}{\partial\sigma}$$
$$\delta\sigma_i^{I-II} = \frac{1}{2}(\delta\sigma_i^{II} - \delta\sigma_i^I)$$
$$\breve{\sigma}_i = \sigma_{i-1} + \frac{1}{2}(\delta\sigma_i^I + \delta\sigma_i^{II})$$
$$\breve{W}_{p,i} = W_{p,i-1} + \frac{1}{2}(\delta W_{p,i}^I + \delta W_{p,i}^{II})$$
$$\xi_i = \frac{\|\delta\sigma^{I-II}\|}{\|\breve{\sigma}_i\|}$$
$$If\ \xi_i > \epsilon\ then\ q = \max\left[0.9\left(\frac{\epsilon}{\xi}\right)^{\frac{1}{2}}, 0.01\right];\ \delta\varepsilon_i = q\delta\varepsilon_i$$
$$Until\ \xi_i \le \epsilon\ (tolerance)$$
$$q = \min\left[0.9\left(\frac{\epsilon}{\xi}\right)^{\frac{1}{2}}, 2\right]$$
$$Stop\ strain\ incrementation\ when \sum_{i=1}^{n} \delta\varepsilon_i = \Delta\varepsilon$$
$$Final\ state\ \breve{\sigma}_i, \breve{W}_{p,i}, C^{ep}(\breve{\sigma}_i, \breve{W}_{p,i})$$

Fig. 1. Modified forward Euler scheme with error control (adapted from Jakobsen and Lade 2002)

The same numerical analysis for the construction phase was reproduced in the present research using a finite difference model consisting of 3212 hexahedral zones in FLAC3D (Fig. 3), all of them with unit thickness and prescribing zero displacement along the longitudinal axis of the dam in order to enable comparison with Kovacevic's results obtained in a 2D numerical analysis. The vertical displacements along lines S1 and S2 are indicated in Fig. 4, showing good agreement between both computed settlement distribution, although differences regarding the field and numerical data may be observed, especially along the S1 line.

Table 1. Single hardening model parameters for loose Sacramento river sand (Lade 2005)

Model component	Parameters	Parameter values		
Elastic properties	M, λ, ν	510	0.28	0.20
Failure criterion	m, η_1, a	0.093	28	0
Plastic potential	ψ_2, μ	−3.72	2.36	–
Yield criterion	h, α	0.534	0.794	–
Hardening law	C, p	0.000127	1.65	–

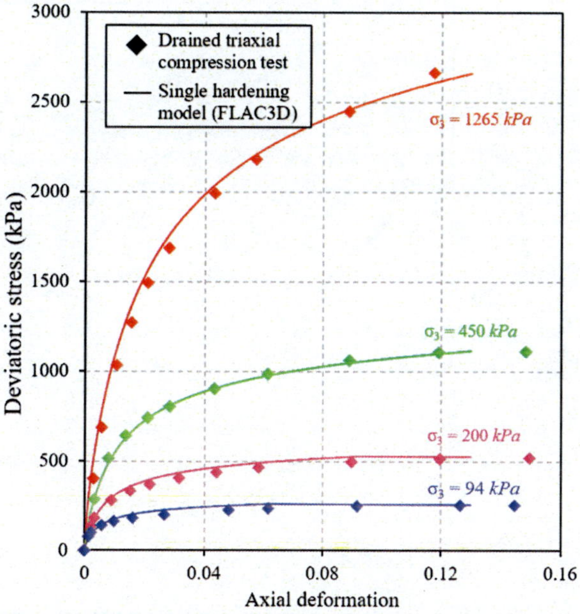

Fig. 2. Numerical simulations of drained triaxial compression tests on loose Sacramento river sand

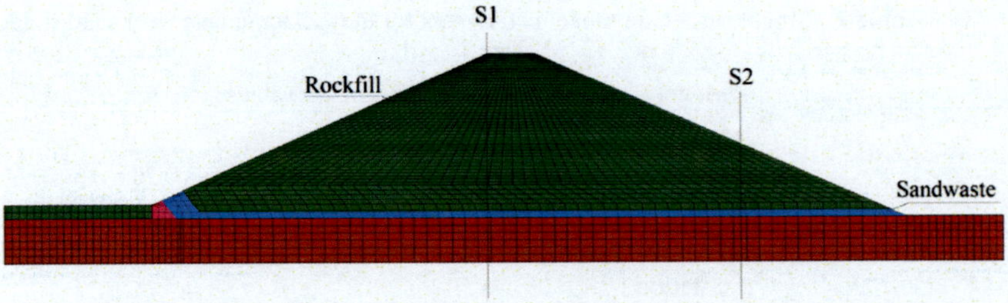

Fig. 3. Finite difference model and location of settlement gauges S1 and S2

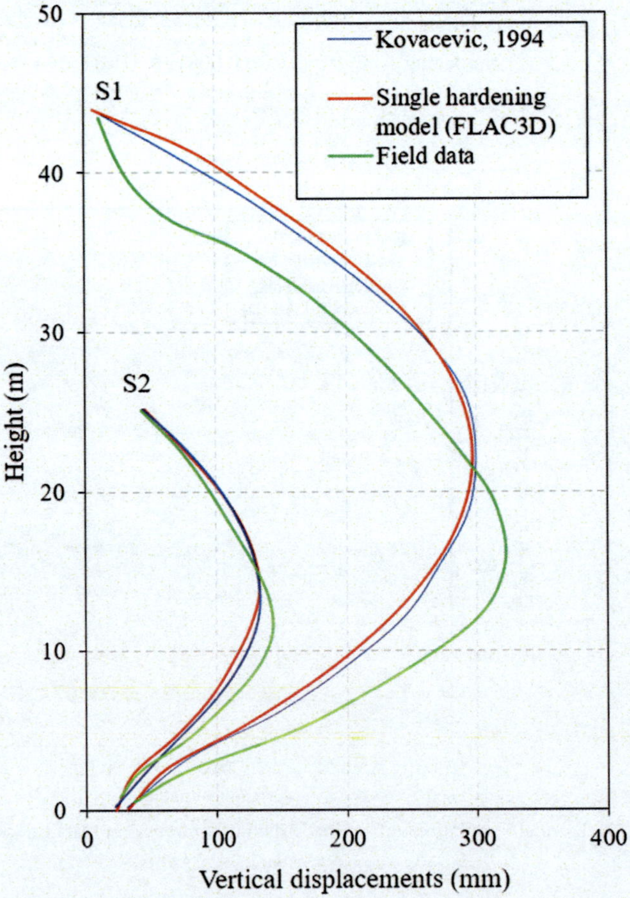

Fig. 4. Vertical displacements along settlement gauges *S*1 and *S*2

Table 2. Single hardening model parameters for rockfill and sandwaste (Kovacevic, 1994)

Parameter	Rockfill			Sand waste		
K_{ur}, n, ν	300	0.50	0.20	1200	0.45	0.20
m, η_1	0.65	260	–	0.10	45	–
ψ_2, μ	−3.00	2.70	–	−3.40	2.15	–
h, α	0.875	0.175	–	0.65	0.45	–
C, p	0.000105	1.80	–	0.00009	1.65	–

4 Case Study – Llancopi Earth Dam, Peru

The Llancopi dam is a homogeneous earth dam located in Cuzco, Peru, constructed with clayey sand (Young's modulus $E = 112.27$ MPa, Poisson's ratio $\nu = 0.2$, friction angle $\phi = 23°$, cohesion $c = 138$ kPa). It is a V-shaped structure with an average height of 16 m and a 220 m long crest. The geometry of the dam was discretized by a 3D finite difference model consisted of 130900 tetrahedral zones, as shown in Fig. 5. The bottom boundary conditions restricted displacements along all three Cartesian axes, while on the lateral faces the displacements were imposed as zero along the respective normal directions. Bulk unit weights were $\gamma = 22.5$ kN/m^3 for the body of the dam and $\gamma = 27$ kN/m^3 for the foundation material. The single hardening model parameters shown in Table 3 were adopted for the analyses, since they allowed to reproduce a stress-strain curve from a conventional triaxial compression test ($\sigma_3 = 135$ kPa) that matched the failure strength obtained with Mohr-Coulomb parameters (Fig. 6).

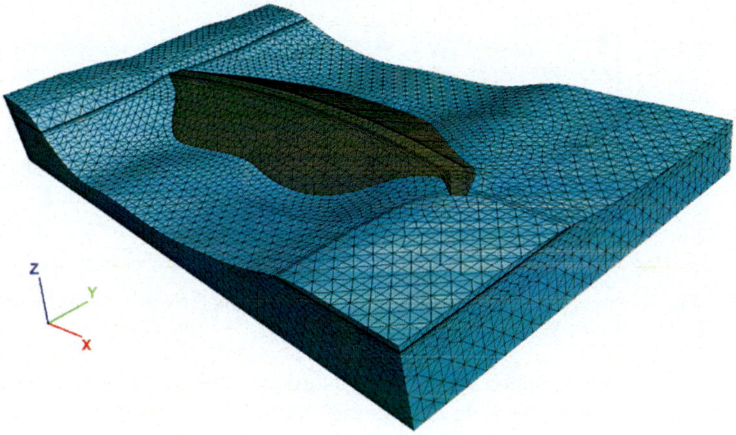

Fig. 5. Three-dimensional finite difference model of the Llancopi dam (Cuzco, Peru)

Table 3. Single hardening model parameters for dense Sacramento river sand (Lade 2005)

Model component	Parameters	Parameter values		
Elastic properties	M, λ, ν	900	0.28	0.20
Failure criterion	m, η_1, a	0.23	80	–
Plastic potential	ψ_2, μ	−3.09	2.01	–
Yield criterion	h, α	0.765	0.229	–
Hardening law	C, p	0.0000396	1.92	–

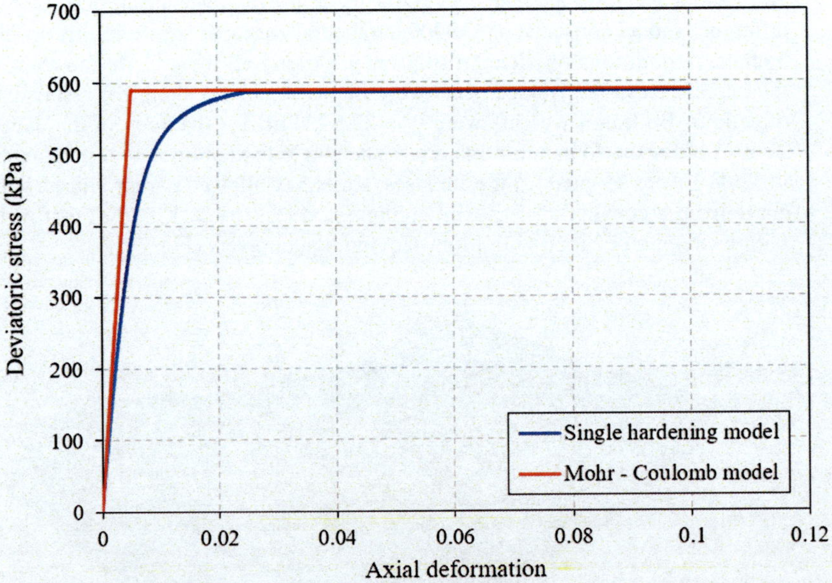

Fig. 6. Stress-strain curves obtained from simulating drained triaxial compression tests

The dam construction was simulated through modeling of successive soil layers and the numerical analyses were done considering both a 3D and 2D geometric configurations, since in the usual 2D plane strain analysis the results may have strong dependency on the cross section chosen as representative of the whole domain.

For the 3D analysis, Figs. 7 and 8 show the distribution of the vertical displacement field at the cross sections at x = 79.5 m and x = 107.5 m indicated in Fig. 9. The settlement profiles along the cross-section centerlines are presented in Fig. 10, where the results obtained in the 2D (plane strain condition) and 3D models are also compared. It is possible to observe some differences for the cross section at x = 79.5 m but, in general, the results computed in both analyses agreed satisfactorily.

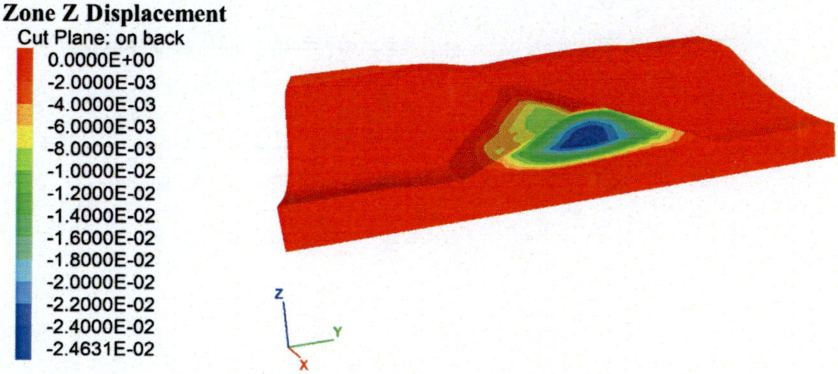

Fig. 7. Vertical displacement field (m) at x = 79.5 *m*

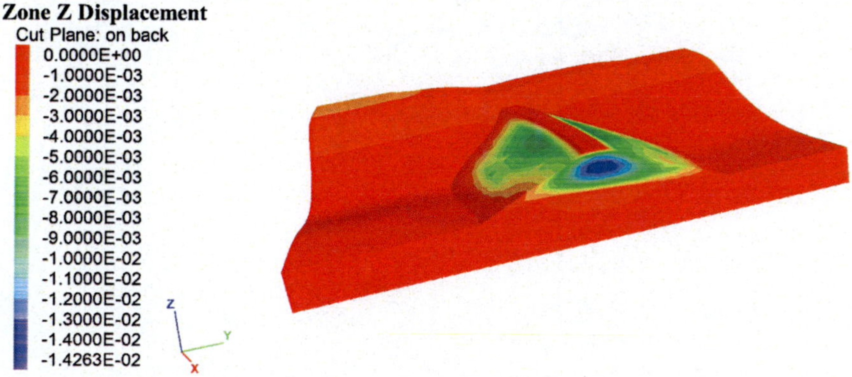

Fig. 8. Vertical displacement field (m) at x = 107.5 *m*

A comparison between 3D predictions using the traditional Mohr-Coulomb model and single hardening constitutive model is shown in Fig. 11 for both cross sections. Much lower vertical displacements were determined when computations were made with the elastoplastic Mohr-Coulomb model, since the soil stiffness before reaching failure is unsatisfactorily represented as it is function of the elastic parameters only.

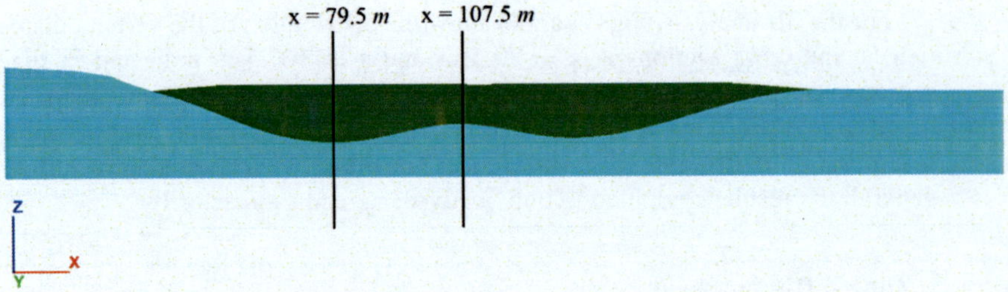

Fig. 9. Positions of the selected cross sections in 2D numerical analyses

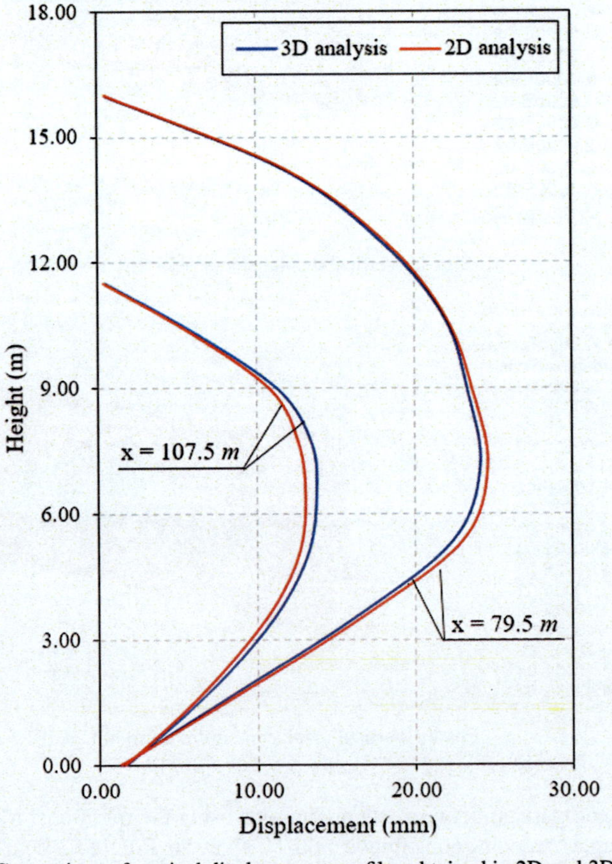

Fig. 10. Comparison of vertical displacement profiles obtained in 2D and 3D analyses

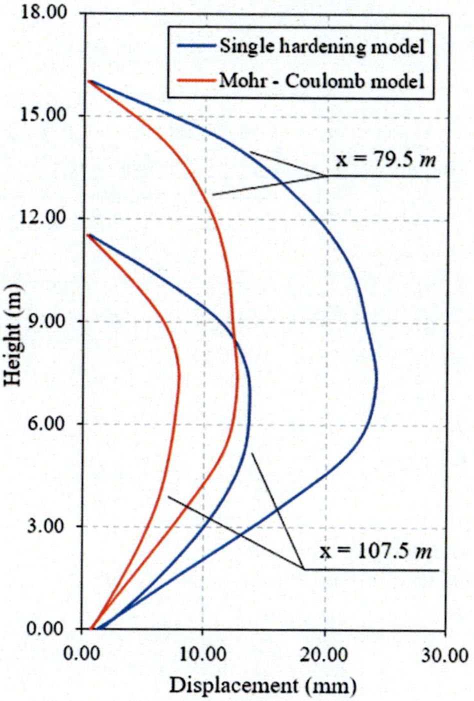

Fig. 11. Comparison of vertical displacement profiles obtained in 3D analysis with the Mohr-Coulomb and the single hardening models

5 Conclusion

The single hardening model (Lade and Kim 1988; Kim and Lade 1988) was implemented in the finite difference program FLAC3D based on the incremental equations presented by Jakobsen and Lade (2002). The validation of the computer code was ensured comparing its numerical results with those previously published in the literature. The Llancopi earth dam, situated in Peru, was then analyzed considering 3D and 2D (plane strain condition) geometries to investigate possible differences in the vertical displacement field in some cross sections of the dam, since engineering projects are usually designed considering just 2D models. Another aspect studied in this research was related to the displacement fields obtained with the advanced single hardening model and the traditional Mohr-Coulomb model where, one time more, it was observed that the Mohr-Coulomb model is inadequate to apply in situations involving stress vs. strain responses of soils.

Acknowledgments. The authors would like to thank the Civil and Environmental Engineering Department at Pontifical University Catholic of Rio de Janeiro for supporting this research.

Appendix

Derivatives of the plastic potential
The derivatives are computed considering the chain rule as follows:

$$\frac{\partial g_p}{\partial \sigma} = \frac{\partial g_p}{\partial I_1}\frac{\partial I_1}{\partial \sigma} + \frac{\partial g_p}{\partial \sigma_2}\frac{\partial I_2}{\partial \sigma} + \frac{\partial g_p}{\partial I_3}\frac{\partial I_3}{\partial \sigma}$$

where

$$\frac{\partial g_p}{\partial I_1} = \left(\psi_1(\mu+3)\frac{I_1^2}{I_3} - (\mu+2)\frac{I_1}{I_2} + \frac{\mu\psi_2}{I_1}\right)\left(\frac{I_1}{p_a}\right)^{\mu}$$

$$\frac{\partial g_p}{\partial I_2} = \frac{I_1^2}{I_2^2}\left(\frac{I_1}{p_a}\right)^{\mu}$$

$$\frac{\partial g_p}{\partial I_3} = -\psi_1\frac{I_1^3}{I_3^2}\left(\frac{I_1}{p_a}\right)^{\mu}$$

Derivatives of the yield function
The derivatives of the yield function are obtained in a similar manner:

$$\frac{\partial f_p'}{\partial \sigma} = \frac{\partial f_p'}{\partial I_1}\frac{\partial I_1}{\partial \sigma} + \frac{\partial f_p'}{\partial \sigma_2}\frac{\partial I_2}{\partial \sigma} + \frac{\partial f_p'}{\partial I_3}\frac{\partial I_3}{\partial \sigma}$$

where

$$\frac{\partial f_p'}{\partial I_1} = \left(\frac{3+h}{I_1} + \frac{\partial q}{\partial I_1}\right)f_p' + \frac{I_1}{I_2}\left(\frac{I_1}{p_a}\right)^{h} e^{q}$$

$$\frac{\partial f_p'}{\partial I_2} = \frac{I_1^2}{I_2^2}\left(\frac{I_1}{p_a}\right)^{h} e^{q}$$

$$\frac{\partial f_p'}{\partial I_3} = f_p'\frac{\partial q}{\partial I_3} - \psi_1\frac{I_1^3}{I_3^2}\left(\frac{I_1}{p_a}\right)^{h} e^{q}$$

in which

$$\frac{\partial q}{\partial I_1} = \frac{\alpha}{\eta_1(1-(1-\alpha)S)^2}\left(\frac{mS\eta_1}{I_1} + \frac{3I_1^2}{I_3}\left(\frac{I_1}{p_a}\right)^{m}\right)$$

$$\frac{\partial q}{\partial I_3} = \frac{\alpha}{\eta_1(1-(1-\alpha)S)^2}\frac{I_1^3}{I_3^2}\left(\frac{I_1}{p_a}\right)^{m}$$

Derivatives of the hardening modulus

$$\frac{\partial f_p}{\partial W_p} = \frac{\partial f_p''}{\partial W_p} = \frac{1}{\rho(Dp_a)^{1/\rho}}W_p^{1/\rho-1}$$

References

Duncan, J.M., Chang, Y.C.: Nonlinear analysis of stress and strain in soils. J. Soil Mech. Found. Div. ASCE **96**(SM5), 1629–1653 (1970)

Jakobsen, K.P., Lade, P.V.: Implementation algorithm for a single hardening constitutive model for frictional materials. Int. J. Numer. Anal. Methods Geomechanics **26**(7), 661–681 (2002)

Kim, M.K., Lade, P.V.: Single hardening constitutive model for frictional materials: I. Plastic potential function. Comput. Geotechn. **5**(4), 307–324 (1988)

Kovacevic, N.: Numerical analyses of rockfill dams, cut slopes and road embankments. Ph.D. Thesis, Faculty of Engineering, University of London, p. 317 (1994)

Lade, P.V.: Overview of constitutive models for soils. In: Soil Constitutive Models: Evaluation, Selection, and Calibration, pp. 1–34 (2005)

Lade, P. V., Kim, M.K.: Single hardening constitutive model for frictional materials II. Yield criterion and plastic work contours. Comput. Geotechnics, **6**(1), 13–29 (1988a)

Lade, P.V., Kim, M.K.: Single hardening constitutive model for frictional materials III. Comparisons with experimental data. Comput. Geotechnics, **6**(1), 31–47 (1988b)

Lade, P.V., Nelson, R.B.: Modelling the elastic behavior of granular materials. Int. J. Numerical Anal. Methods Geomechanics **11**, 521–542 (1987)

Sloan, S.W.: Substepping schemes for the numerical integration of elastoplastic stress–strain relations. Int. J. Numerical Methods Eng. **24**(5), 893–911 (1987)

Wyllie, C.R., Barrett, L.C.: Advanced Engineering Calculus, McGraw-Hill (1982)

Wilson, A.C., Evans, J.D.: The use of low grade rockfill at Roadford Dam. In: The Embankment Dam: Proceedings of the Sixth Conference of the British Dam Society held in Nottingham on 12–15 September 1990, pp. 21–27. Thomas Telford Publishing (1990)

Geomembranes to Line Surge Shafts

Alberto Scuero[1] and Gabriella Vaschetti[2]

[1] Carpi Group, Balerna, Switzerland
alberto.scuero@carpitech.com
[2] Carpi Tech, Balerna, Switzerland
gabriella.vaschetti@carpitech.com

Abstract. Surge shafts are underground water structures that are usually lined. The lining provides stability and watertightness to the excavation line, avoiding that excessive water infiltration undermines the stability of the slopes and of the structure itself. Traditional lining systems have been reinforced or un-reinforced concrete and steel. The construction of concrete and steel linings is expensive and time consuming, it can be further complicated in locations with difficult access, and it may require expensive maintenance. Over time, or in presence of particularly difficult geological conditions, concrete linings deteriorate, water infiltrates into the soil through fissures, or/and failing joints and increased permeability, increased roughness results in reduced hydraulic efficiency. Similarly, steel liners may be stressed above their yielding point. Loss of watertightness can eventually trigger landslides and ultimately cause partial or total collapse of the structure. Waterproof geomembrane lining systems based on the use of thermoplastic materials can efficiently substitute traditional linings for rehabilitation as well as for new construction. They can be installed in short times, and are a permanent solution to provide watertightness, reduce head losses, and maintain the shaft in its design conditions. The high flexibility and elongation capability of the thermoplastic geomembrane allows bridging construction joints and existing cracks and accommodating the opening of new cracks. In case of difficult geology, a geomembrane system can increase the safety of the shaft by maintaining the watertightness for which it has been designed. The paper discusses the design concepts and components of drained geomembrane lining systems anchored by longitudinal fastening lines and by watertight perimeter seals, through case histories of surge shafts projects recently completed, one in South America and one in the middle East. In both projects the geomembrane system was installed on the existing concrete lining and left exposed to the water.

1 Introduction

This paper discusses how exposed geomembrane systems can provide or restore watertightness in surge shafts. The rationale is the same conceived and adopted in the 1980s in pioneer projects in hydro tunnels, and further developed in the 1990s for high-pressure tunnels, such as Thissavros tunnel in Greece (1997), inside water pressure 1.5 MPa, maximum water velocity exceeding 4 m/s, negative uplift pressure about 0.5 MPa. Recent projects in pressure tunnels can accept much higher heads: for Helms Penstock Access

H. Shehata and M. Badr (Eds.): *Advancements in Geotechnical Engineering*, SUCI, pp. 106–120, 2021.
https://doi.org/10.1007/978-3-030-62908-3_8

Tunnel (USA 2017), approximately 76 m long and >8 m in diameter, terminating into a 10.6 m thick concrete plug with maximum static hydrostatic pressure exceeding 5.1 MPa, the geomembrane system has been tested under the required maximum pressure of 8.52 MPa.

After discussing the rationale of the state-of-the-art system, the paper presents two recent case histories: Tunjita shaft in Colombia, 2017, and Gilboa shaft in Israel, 2018.

2 Background

Hydraulic tunnels and shafts are underground water ways that, with the exception of old tunnels excavated in rocks, are usually lined. The main function of the liner is to provide stability and watertightness to the excavation line, avoiding that excessive water infiltration may undermine the stability of the slopes and of the structure itself. The liner also aims to minimise head losses, thanks to its low friction.

Traditional lining systems have been reinforced or un-reinforced concrete and steel. Besides the fact that the construction of concrete and steel linings is expensive and time consuming and can be further complicated in locations with difficult access, the main drawback of concrete linings is that over time they deteriorate: fissures, failing joints and increased permeability lead to water infiltration into the soil, increased roughness results in reduced hydraulic efficiency. In presence of particularly difficult geological conditions, concrete and steel liners may experience problems due to excessive stresses arising for example from ground movements, which may exceed the strength of the concrete or the yielding point of the steel, and lead to failure of the liner.

Deterioration of the lining and subsequent loss of watertightness may have serious consequences: continuous water seepage into the surrounding soil may in the long term affect the stability of steep natural slopes, eventually triggering landslides that can ultimately cause the partial or total collapse of the structure. The owners therefore face risk of instability and power loss, besides head loss due to increased roughness.

To restore watertightness and bring the pressure tunnel/shaft back to its design conditions, expensive outage is required, therefore rehabilitation measures must be carried out in the shortest possible time. Local repairs with mortars and resins have shown drawbacks in respect to efficiency and durability. Overall repair with a new liner must be designed to optimise several parameters: hydraulic efficiency, lifetime, outage time for installation, costs, future maintenance. Geomembrane systems, based on the use of practically watertight thermoplastic materials, have proved to be a permanent repair solution highly efficient from a technical, construction and operation perspective.

This paper focuses on exposed geomembrane systems based on the use of SIBELON® composite geomembranes (geocomposites), which have proven to be able to provide long-term watertightness, both in rehabilitation and in new construction. In new construction, implementing the concrete layer with such a geomembrane liner since the design stage will increase the project's safety, because even if the concrete cracks due to large or unforeseen movements in the natural ground, including seismic events, the geomembrane will maintain the efficiency of the water barrier. Additionally, the low hydraulic roughness that the geomembrane will maintain over time will minimise head losses and ensure long-term hydraulic efficiency.

3 SIBELON® Geomembrane Systems in Pressure Tunnels and Shafts

3.1 Design Concepts

The rationale of the system is to provide over the entire cross section of the structure a practically watertight liner capable to maintain imperviousness under the acting loads, and with elongation characteristics allowing bridging those discontinuities that may arise in the subgrade (e.g. cracks forming in the concrete liner on which the geomembrane is applied). The geomembrane barrier prevents water entering the deteriorated concrete lining through cracks and joints, and affecting the durability of the reinforcement, and prevents water escaping into the surrounding ground with the potentially dangerous effects described. Additional assets are that the exposed geomembrane system does not significantly reduce the tunnel/shaft envelope, and the low hydraulic roughness that the geomembrane will maintain over time will improve hydraulic efficiency and minimise head losses. Since no routine maintenance is required, costs of future maintenance and related outage costs are deleted or minimised.

Design parameters for the choice of the liner, of its anchorage system, of drainage and support layers if any, are the water head, the static and dynamic pressure, the water velocity, the negative pressure outside the structure when it is empty, the way of operating the structure, the stability and roughness of the subgrade, etc. The geomembrane liner is anchored by mechanical fastening lines whose type, number and position are designed based on the mentioned design loads at full and empty structure, and taking into account also the dynamic action of water that would result from an accidental damage in the geomembrane liner that allows water flowing under it. Water velocities of several meters/second and pressures of several MPa can be accommodated.

A drainage system is provided behind the geomembrane to collect and discharge water that should be present behind the geomembrane liner, be it water migrating from the slopes towards the structure, or already impregnating a deteriorated concrete liner. The drainage system avoids that if the structure is dewatered backpressure can build up and ultimately exceed the resistance to burst of the geomembrane.

3.2 Components

The liner adopted in all state-of-the-art projects is SIBELON® CNT, a geocomposite consisting of a flexible polyvinylchloride geomembrane plasticised with high molecular weight branched plasticisers, and heat-bonded at fabrication to an anti-puncture geotextile. The geomembrane, with hydraulic conductivity $k = 6.25 * 10^{-14}$ cm/s, provides the water barrier, the geotextile protects the geomembrane against puncturing from the substrate. The design strength of the geocomposite liner is assumed to be the geotextile break tension, in the range of 50–60 kN/m and corresponding to a strain $>60\%$, while the geomembrane characteristic elastic behaviour up to a strain $>250\%$ is activated only in case excessive stresses would produce delamination of the backing geotextile. For design purpose and in favour of safety, the strength parameters of the geocomposite are corrected to account for potential minor defects not detected by the QC process and for long-term ageing of the polymeric material. The thickness of the geomembrane and the mass per unit area of the geotextile are selected based on the specific project's requirements.

The face anchorage system consists of continuous stainless-steel anchorage lines parallel to the water flow. The objective of these longitudinal fastening lines is to keep the liner stable and taut to the surface. This objective can be achieved either by a couple of tensioning profiles, or by batten strips: the tensioning profiles avoid slack areas and folds by exploiting their particular geometry and the way they are connected, the batten strips by the way they are installed, as further detailed. The design of the anchorage lines is based on worst case scenarios, aiming at avoiding that extreme loads may cause the failure of the anchorage system with the risk of detachment of metal elements and geocomposite fragments. The calculations are based assuming that, in case of accidental damage resulting in perforation and/or cutting of the geocomposite liner, the water infiltrating through the damage can generate a negative pressure behind the liner equal to the internal pressure in the tunnel/shaft. In this condition, remote but crucial for a correct design, the kinetic head of the water flow can generate uplift and deformation of the geocomposite liner. The tension in the geocomposite is obtained by an equation based on the kinetic head (function of the water velocity and density) and on the geometry of the deformation, and using the stress-strain curve of the geocomposite. The tension in the geocomposite liner thus obtained must be at least equal to what imposed by the established Factor of Safety (FS). The tension transmitted to the anchorage lines results in a pull-out component and a shear component: the design condition for the calculation of the pull-out force on the anchors corresponds to the case in which two adjacent sheets of geocomposite liner are simultaneously damaged and uplifted, resulting in the combination of the pull-out components (one for each sheet), while the design condition for the calculation of the shear force on the anchors corresponds to the case in which only a single sheet of geocomposite liner is uplifted, resulting in an asymmetric shear component. The data obtained by calculations of the tension in the geocomposite, of FS on the geocomposite, of pull-out and shear forces on the anchors, and of FS for pull-out and shear, verify the soundness of the anchorage system.

A transverse stainless-steel watertight seal is placed at the beginning and end of the lined area to avoid water infiltrating behind the liner.

The drainage system is designed based on an analysis of the estimated amount of water that can accumulate behind the liner, leaking through the ground/cracked concrete lining when the tunnel/shaft is empty. The inflow is essentially due to a) the intrinsic permeability of the outer ring/concrete lining and b) the leakage through the concrete cracks. Given the low permeability of the concrete, leakage through cracks can be considered as the primary source of inflow, while other sources can usually be neglected. A continuous drainage layer consisting of a geosynthetic having high in-plane transmissivity is often applied to enhance efficiency of drainage transmission to the custom-designed drainage discharge components. In case of subgrade with large cracks/active cracks/large discontinuities/large irregularities, additional geosynthetic layers are used to provide support/anti-puncturing.

3.3 Past Experience and Performance Data

To the knowledge of the authors, experience of watertight geomembranes in high-pressure conduits dates back to the late 1970s- early 1980s, when Austria saw several applications with water heads in the range of 0.2 to 0.7 MPa, where the geomembrane was sandwiched between the outer regularization ring and the inner concrete ring. Carpi started installing totally or partially exposed geomembranes in pressure tunnels in 1984 (Gorghiglio pressure tunnel, Italy). The feedback of several installations proves the effectiveness of the system and its capability to resist high pressures and water velocity. Outstanding examples detailed in international literature are Thissavros (Scuero and Vaschetti, 2005) and Belden (Wilkes et al. 2011).

Thissavros is an 11 m diameter pressure tunnel in Greece, with inside water pressure of 1.5 MPa and maximum water velocity exceeding 4 m/s. Water infiltrating from the reservoir generates a negative uplift pressure of about 0.5 MPa. The tunnel was originally lined with concrete, which started to crack. At first reservoir filling, because of some changes in the location of the dam grout curtain and of the cracks in the concrete of the tunnel, the dam galleries were flooded. It was decided to extend upstream the concrete plug for a length of 60 m. Time constraints and a much more attractive price of an exposed geomembrane system suggested to split the required extra works in two sections: one section extending the conventional concrete plug and its steel lining, the second section using only an exposed SIBELON® CNT geocomposite system upon the fractured concrete liner. Installation of the geomembrane system, in 1997, required 3 weeks and the cost was approximately 25% of the cost of an equivalent steel liner, which would have required more than 3 months' time for installation. A few weeks after the system had been installed and the tunnel had started operating, the tunnel had to be emptied to execute some extra grouting works in correspondence of the concrete plug. That situation allowed testing the fully satisfactory behaviour of the geocomposite and of its drainage system, which worked as designed. After grouting works were completed, the tunnel was filled again and it will be always full of water, with no plan to dewater it again. The geomembrane liner is working effectively since 1997.

Belden hydropower plant, built in 1958 in USA, includes a 2941.41 m long concrete lined tunnel (Tunnel 2). In 1970, during an outage, two large circumferential cracks and additional cracks were identified in the lining of Tunnel 2. The cracks were repaired on several occasions; instrumentation showed that the ground was continuously moving and slowly accelerating, leakage from the tunnel being the main cause for the sliding of the ground. The studies undertaken by the owner to determine the best method to rehabilitate the tunnel eventually lead to the selection of an exposed geomembrane liner because of speed of installation, lower cost, and potential ability to accommodate future ground movements. The geomembrane design was challenging due to the changing shape of tunnel, the water velocity (4.27 m/s), the hydraulic head (0.54 MPa), the groundwater pressure (0.24 MPa), and the drainage requirements due to the high water table in the mountain and the need to drain upstream. A complete survey of the tunnel showed that in one section the cracks could reach width >22 mm over the design life of the project. Such opening, together with the pressure applied to the surface, could induce large strains in the selected SIBELON® CNT geocomposite. A support system able to withstand the load transmitted by the water pressure on the span created by the big cracks was designed. Installation was made in 2008, the tunnel was immediately put back to operation for a couple of days, then dewatered and inspected, and put back in service. In 2014, 6 years after installation, the tunnel was again dewatered and inspected. The detailed inspection did not find one single damage, no repairs were needed, and the tunnel was put back in service. Data readings have shown that the geomembrane system is successfully dewatering the mountain and preventing seepage that could aggravate ground movement.

4 Tunjita Surge Chamber and Shaft

Project Data

A recent example of application on a shaft is at Tunjita power plant, in Colombia. The small Tunjita scheme includes a surge chamber, a shaft, an access gallery at the bottom of the shaft and a headrace tunnel (Fig. 1). All these structures are lined with three 0.05 m thick layers of shotcrete reinforced with a steel mesh, and with a 0.25 to 0.40 m thick layer of reinforced concrete. The surge chamber is about 71.5 m high (from El. 1611.50 m to El. 1540.00 m), has a circular section with internal diameter 4.1 m, and a maximum hydrostatic pressure of about 0.6 MPa. The shaft is about 108.9 m high (from El. 1540.00 m to El. 1431.04 m), has a circular section with an internal diameter of 2.5 m, and a maximum hydrostatic pressure of about 1.74 MPa. The bottom part of the surge chamber (about 50 m) and the whole shaft are permanently below the water table, with consequent presence of infiltration water from the rocky mass when the structures are empty.

During a routine inspection conducted at beginning of 2017 it was found that the existing concrete lining in the surge chamber, shaft and pressure tunnel was heavily affected by cracking, with some parts of the lining detaching. A safety water barrier was deemed necessary for these structures, to prevent leakage through the cracks and avoid further deterioration of the concrete liner, so that the scheme could continue operating safely. The selected water barrier, which besides its technical advantages could be installed in a short time profiting of the outage period, was a SIBELON® geomembrane system installed over the deteriorated concrete, left exposed to the flowing water and mechanically anchored. The geomembrane system was installed from El. 1601 (a little lower than the top of the surge chamber) down to the bottom of the shaft (about El. 1431), and continuing into the elbow and the pressure tunnel.

Waterproofing System

The conditions of the deteriorated concrete liner (large cracks, irregularities) and the high water pressure required supporting the waterproofing geocomposite to avoid it being forced into the cracks or punctured on the rough subgrade by the water head. A geocomposite formed by a geogrid with bidirectional polypropylene mesh with high bearing resistance and high modulus, laminated during manufacturing to a non-woven polypropylene anti-puncture geotextile, was installed as support layer over the deteriorated existing concrete liner (Fig. 2).

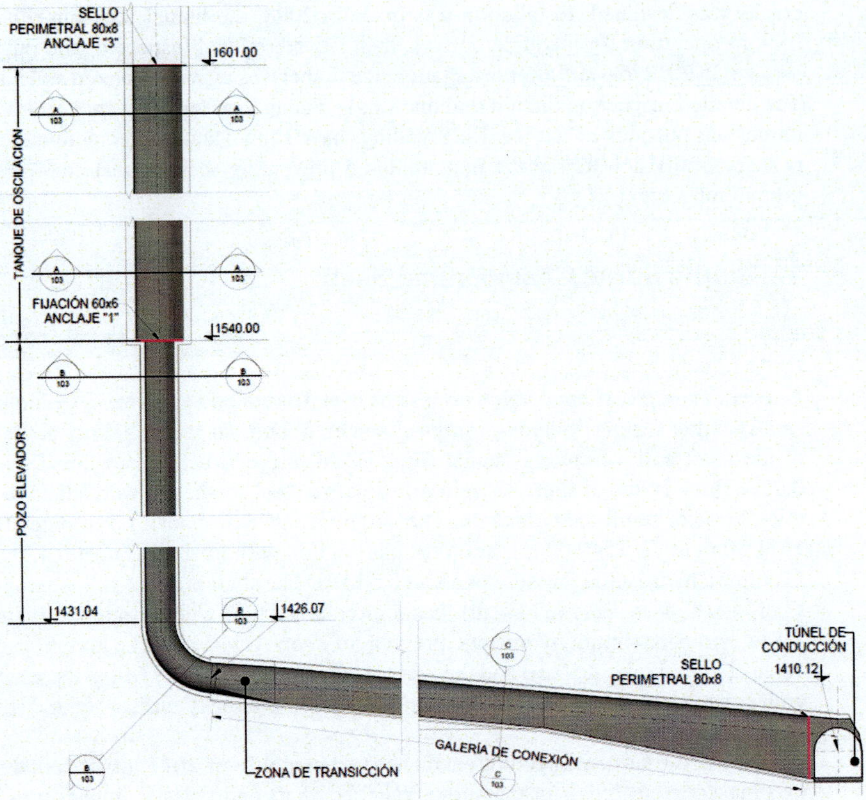

Fig. 1. The Tunjita scheme

Fig. 2. Tunjita. The deteriorated concrete lining in the shaft, and vertical view from bottom to top of the support geocomposite

The selected waterproofing liner is the geocomposite SIBELON® CNT 4400 (3.0 mm thick geomembrane heat-bonded to a 500 g/m^2 geotextile). The geocomposite sheets, 2.10 m wide, with length appropriate to span the section to be lined, were temporarily anchored at top and deployed vertically down. Adjacent sheets were joined vertically by watertight manual heat-seams, all checked according to ASTM standards.

For the first time in the world in a water conveyance structure, a system different from the standard tensioning system was adopted for the longitudinal fastening of the waterproofing liner. The fastening system consists of longitudinal stainless-steel batten strips, cross section 60 × 6 mm, anchored by mechanical anchors selected in function of the strength of the concrete. Tensioning of the geocomposite is achieved by an installation sequence based on the concept of "progressive closure", whereby the fastening lines are not all placed at the same time: first are placed the "first phase" anchorage lines, and only after they have been permanently anchored the "second phase" anchorage lines are placed and permanently anchored. This progressive installation method produces a tensioning effect on the geocomposite liner in between two adjacent first phase anchorage lines, preventing formation of slack areas and wrinkles in the geocomposite (Fig. 3 at left, and Fig. 7 related to Gilboa shaft). The water pressure will further push the geocomposite liner to adhere against the concrete lining, with an additional tensioning effect. The number of fastening lines, calculated with the same rationale described further on for the Gilboa scheme, amounts to 10 for the surge chamber, and 8 for the shaft.

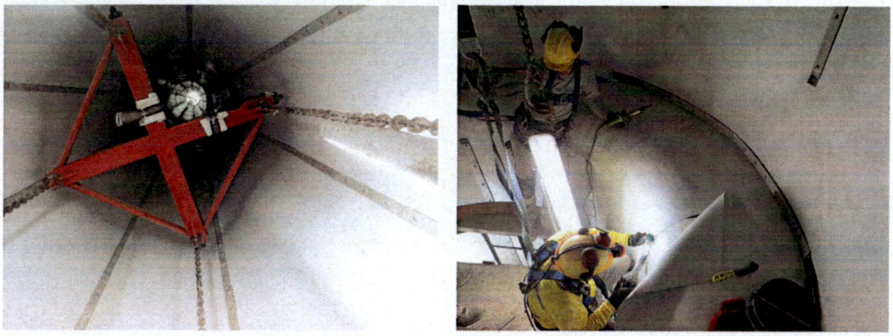

Fig. 3. Tunjita. The waterproofing geocomposite fastened and tensioned, and the perimeter seal at the transition surge chamber/shaft

Watertight perimeter seals confine the waterproofing liner at top and bottom of the surge chamber (El. 1601 and El. 1540). The seals consist of stainless-steel batten strips bolted to the concrete by mechanical anchors at 15 cm spacing, and with a cross section of 80 × 8 mm at top and 60 × 6 mm at bottom (Fig. 3 at right) at the bottom of the shaft; the waterproofing liner continues into the pressure tunnel.

Waterproofing works were executed from a travelling platform suspended at a steel structure outside the top of the surge chamber. The same platform was used to lower all materials for lining the surge chamber and shaft to the staging area. Installation was carried out from August to October 2017. In total, the area lined with the geocomposite is 1,878 m^2.

5 Gilboa Shaft

Project Data

The Gilboa pumped storage scheme in Israel, installed capacity 300 MW, comprises an underground powerhouse, an approximately 480 m deep high-pressure shaft with 4.5 m constant internal diameter, followed by an approximately 900 m long high-pressure tunnel with 4.0 m constant internal diameter, and a low-pressure tunnel (Fig. 4). The shaft and the high-pressure tunnel are separated by an approximately 8 m radius elbow. The entire shaft is concrete lined, whereas the high-pressure tunnel comprises a first concrete lined section approximately 325 m long and a second steel lined section approximately 575 m long (up to the penstock bifurcation).

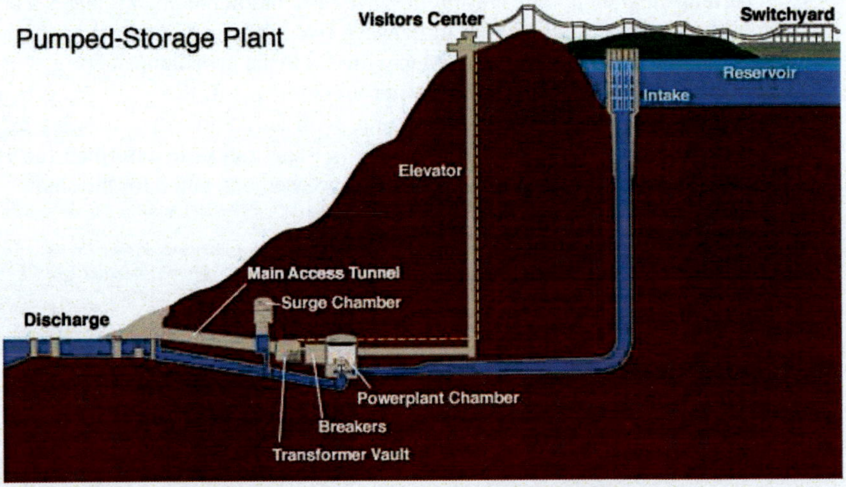

Fig. 4. General scheme of the Gilboa PSP

The 485 m deep shaft was built with the "raise boring" method while the tunnel was built with the "drill and blast" method. In the shaft, the rock excavation has been stabilized by a nearly 0.1 m thick shotcrete layer on which a heavily reinforced 0.5 m thick concrete lining has been placed, whose thickness in the bottom section between El. −27 m and −54 m is increased to 1.5 m. The section below El. −86.5 m includes a 4.5 m long transition with internal diameter from 4.5 m to 4.0 m protected with a metal lining. The steel reinforcement varies along the shaft according to the properties of the rock mass. Maximum permissible water velocity (concrete lining) is < 6 m/s.

Maximum internal instantaneous pressure at the intake elbow (about El. +389 m) is 0.35 MPa, pressure at the bottom elbow (about El. −91 m) is about 5.94 MPa. Ground water level is at El. −61 m. Maximum external (negative) pressure is potentially equal to the internal static pressure due to the leakage through the concrete lining. The concrete surface presents, especially close to the bottom elbow, defects such as honeycombs, small horizontal cracks, small holes (Fig. 5 at left). Some defects have been repaired with resins; results of such local repairs have been unsatisfactory (detaching resin, Fig. 5 at right).

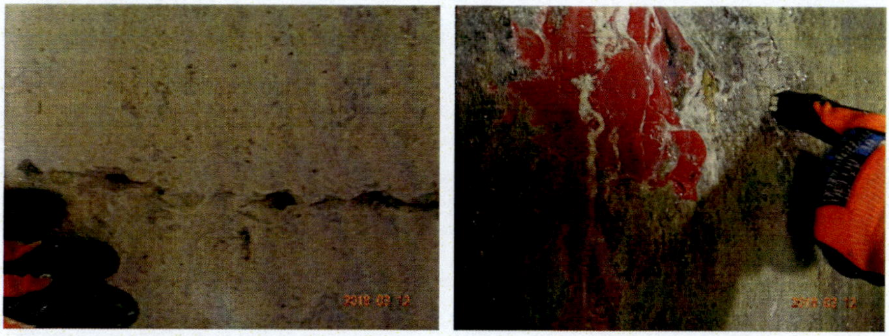

Fig. 5. Gilboa Shaft. Defects in the concrete surface. On right, honeycombs repaired with resin, which is detaching and causing leakage to reoccur

Since potential leakage through the concrete lining was expected, an exposed geomembrane system was selected as safety water barrier to the shaft and to about the first 376 meters of the high-pressure tunnel, to prevent leakage and loss of valuable water. The shaft was waterproofed for the full-round cross-section over its entire height, starting at the top at El. +389 m down to the beginning of the transition protected with a metal lining at about El. −89.5 m (just above the bottom bend), for a height of nearly 479 m. Figure 6 shows the lined area and the anchorage system.

Design Parameters and Issues

The calculation of the anchorage system assumed that, in case of accidental damage to the geomembrane liner, the leakage of water could generate a negative pressure behind the liner equal to the internal pressure in the shaft. In this unlikely condition, the kinetic head of the water flow can generate an uplift of the liner. The design aimed to prevent exceptional loads from causing failure of the anchorage system, risking detachment of metal elements and geomembrane fragments. The design water velocity was 6 m/s. The minimum Factor of Safety selected by Carpi is FS $\geq$ 1.5. As a conservative measure, a Partial Safety Factor on geomembrane material's strength of 2.0 was introduced. Partial Safety Factors are also included by the manufacturers of anchors.

The capability of the waterproofing liner to elongate on potential cracks in the concrete liner (bridging capacity) was calculated based on the design data, with a specified concrete lining crack opening criterion < 0.2 mm, and a maximum pressure head of 5.9 MPa.

The geomembrane system is drained behind. Drained water discharge occurs inside the shaft, via one-way mechanical valves. The design of the discharge capacity considered an inflow due to intrinsic permeability of the concrete lining and leakage through the concrete cracks, foreseen by the above criterion.

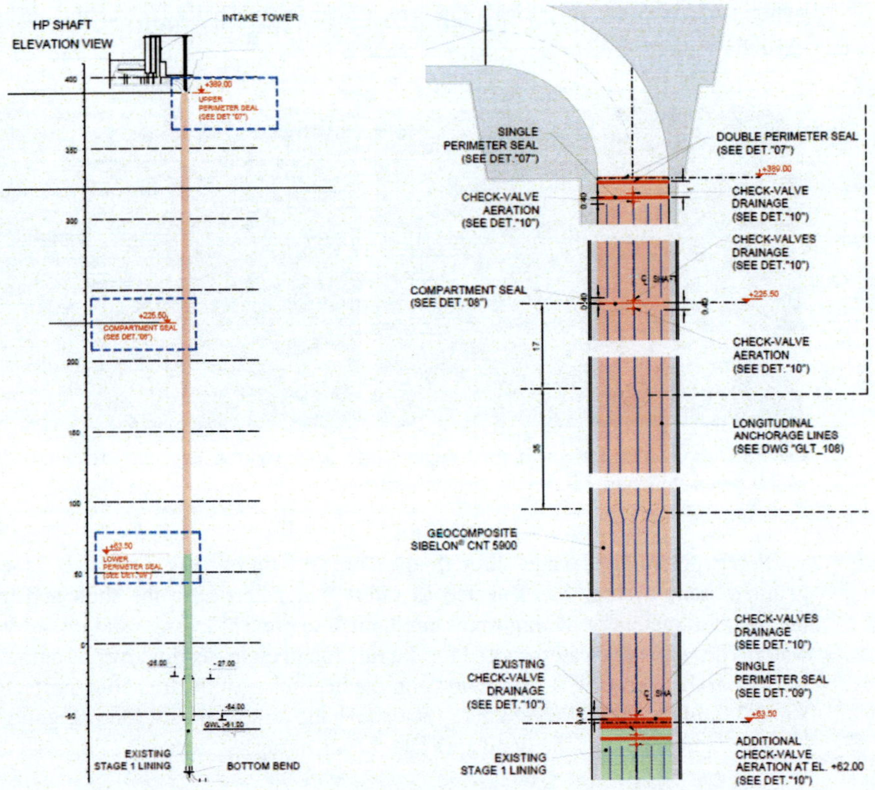

Fig. 6. Gilboa shaft. Scheme of lined area and of anchorage system

The announced steel reinforcement concrete cover was 50 mm with tolerance ± 10 mm, which required using special mechanical anchors for short embedment. During installation, it was found that the concrete cover was thicker (70 to 150 mm).

Waterproofing System

The waterproofing liner from El. −89.5 to El. +63.5, and partially from El. −359 to El. +388, is SIBELON® CNT 4400 (a 3.0 mm thick SIBELON® C geomembrane heat-bonded during fabrication to a 500 g/m^2 non-woven, needle punched, polypropylene geotextile). From El. +63.5 to El. +359 and partially from El. −359 to El. + 389, the waterproofing liner is SIBELON® CNT 5900 (a 4.0 mm thick SIBELON® C geomembrane heat-bonded during fabrication to a 700 g/m^2 non-woven, needle punched, polyester geotextile). The geocomposite is secured to the concrete by 14 fastening lines placed at regular spacing (about 1 m apart). Each fastening line consists of 60 × 6 mm stainless-steel batten strips bolted to the concrete by mechanical anchors at about 20 cm spacing. All fastening lines are intrinsically watertight. As at Tunjita, to avoid slack areas where excessive stresses could be exerted on the geocomposite, a "progressive closure" process was adopted. Figure 7 gives evidence of the tensioning effect obtained by the process. As mentioned for Tunjita, the water pressure will further push the geocomposite liner to adhere against the concrete lining, with an additional tensioning effect.

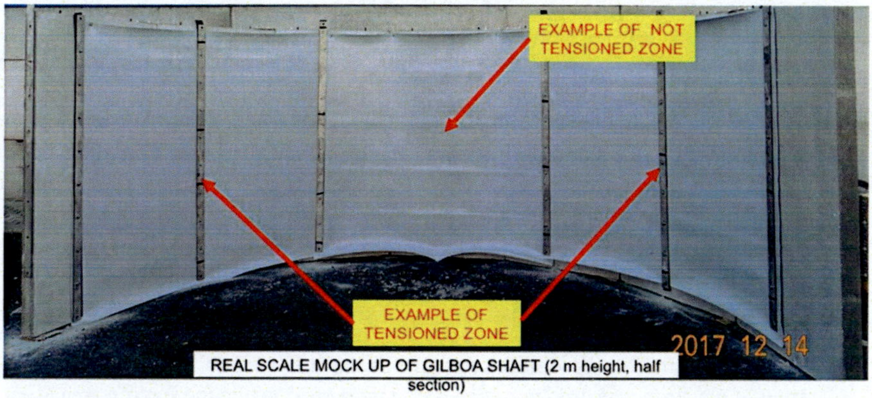

Fig. 7. Gilboa shaft mock-up showing the tensioning concept. In the centre, the "first phase" batten strips are placed at 2 m spacing, no tensioning effect. At the sides, the "second phase" batten strips have been placed, spacing is 1 m, tensioning effect achieved

The waterproofing system is sealed at top and bottom by a triple watertight perimeter seal (Fig. 8): two adjacent seals for increased safety are placed at top and bottom elevations, a third seal is placed at 1 m distance from them. All seals consist of 80 × 8 mm stainless-steel batten strips bolted to the concrete by mechanical anchors at 15 cm spacing.

Fig. 8. Gilboa shaft. At left, the top part of the lined area: the double perimeter seal to provide extra safety, the third seal one metre underneath, and the top of the vertical fastening lines. At right, the same perimeter seals arrangement at the bottom of the lined area, and the one-way valves for relief of backpressure

The drainage system behind the geomembrane consists of the drainage gap allowed by the vertical fastening lines, and of four check-valves (one-way relief valves) placed at the bottom of the lined area, three in the one meter spacing between the seals, and one above the third seal (Fig. 8 at right).

Installation: Challenges and Procedures
One major challenge faced during installation was respecting a tight schedule, especially in the first campaign when the bottom 150 m of the shaft were waterproofed: to ensure that the bottom part with the highest pressure would be waterproof, which was mandatory for the safety of the shaft, performance testing was needed, therefore the waterproofing works had to be performed as quickly as possible, to allow the shaft being put back into operation for testing.

Access to the staging area complicated the time challenge, because going down each day from El. +389 m to the staging area was time consuming. Two travelling platforms of different sizes were used (Fig. 9), suspended at a steel structure outside the top of the shaft. The larger platform, almost the size of the shaft and used to transport materials and to install the waterproofing system, was moved based on material supply needs. The smaller platform was used to transport the workers down to the staging area at each shift. The travelling time of the bigger platform was 20 to 25 min. The travelling time of the smaller platform varied from 10 to 25 min, depending on the depth and on the extent of the inevitable oscillations.

Fig. 9. Gilboa shaft. At left, the small platform descending to the installation platform. At right, the workers on the installation platform

Installation was carried out in separate horizontal stages. Watertight horizontal heat seams connect adjacent stages. Safety standards for underground working environment required personal safety equipment including filtering half masks (Fig. 10). Final operations included perimeter sealing and installation of the one-way relief valves.

Fig. 10. Gilboa shaft. Heat-seaming of adjacent sheets is done manually and checked with ASTM standards. Note the filtering half masks to minimize impact of fumes caused by seaming operations

In the first campaign in spring 2018 the system was installed in the bottom 150 m of the shaft, in total 2,120 m^2 installed in 4 weeks. After completion of the works in the first campaign the TFT (Tunnel Filling Test) was performed by the EPC contractor to verify the performance of the full high-pressure shaft and tunnel. The results weren't satisfactory due to the section of the shaft that was not lined with the geomembrane system. This was the reason why the Owner's Engineer (EDF) required that the full shaft and the concrete lined section of the tunnel be protected by the geomembrane system. In the second campaign in fall 2018 the rest of the shaft was lined, in 48 days, for a total lined area of 6,765 m^2. After completion of the waterproofing works in the entire shaft, a new TFT was executed; the analysis of results of the first and the second filling test showed that the waterproofing system reduced the leakage measured during the first TFT of more than 95% and well below the Employer's requirements (Figs. 11 and 12).

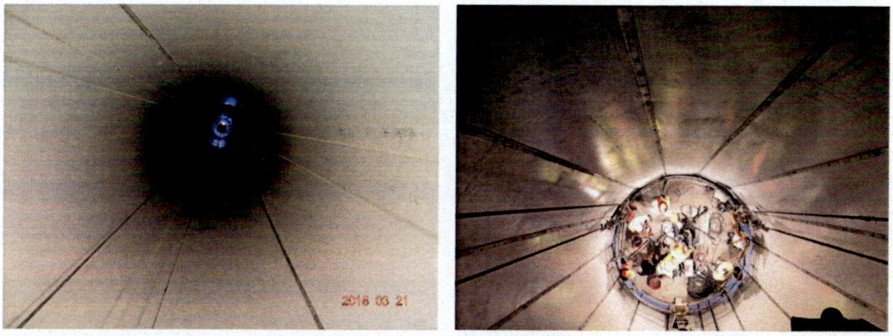

Fig. 11. Gilboa shaft. At left, looking from the bottom of the shaft upwards (depth can be appreciated by the blue hole indicating the shaft top external exit), where only the first phase fastening lines have been placed. At right, looking down from top, second phase fastening lines have been placed, in total 14 fastening lines to resist about 5.9 MPa pressure

Fig. 12. Gilboa shaft. Small platform traveling to surface after work completion

6 Conclusions

The presented case histories are outstanding projects due to the magnitude of the loads involved. Installation challenges were significant and were met successfully through careful equipment planning and work scheduling, maintaining high safety standards.

The vast experience gained at these and at similar projects with high water heads and water speed, along with the availability of high-performance materials, allows developing sound design for such structures.

References

Scuero, A., Vaschetti, G.: Stopping water and power losses in hydraulic tunnels. Techniques of Reinforced Concrete for Hydraulic Engineering (2005)

Wilkes, J., Yu, A., Mcmanus, R., Jaramillo, V.C.: Lining Belden Hydraulic Tunnel with a Geomembrane. World Tunnel Congress (2011)

Expanded Polystyrene in Soil Reinforcement

David Carvalho[1], Roberto Kassouf[2(✉)], Pedro Scatena[3], and Vilson Scatena[4]

[1] State University of Campinas, Campinas, Brazil
david@feagri.unicamp.br
[2] UniMetrocamp Wyden, Campinas, Brazil
robertokassouf@terra.com.br
[3] Termotécnica, Campinas, Brazil
pedro.scatena@termotecnica.com.br
[4] UniMetrocamp, Campinas, Brazil
vilsonscatena@gmail.com

Abstract. The design and execution of works on soft soils represent high complexity and challenges to geotechnicians, since these soils have low bearing capacity. Traditional construction methods require total or partial replacement of low resistance soil by soil with good bearing capacity; sometimes it is possible to construct lateral equilibrium berms; or to construct vertical drains that help accelerate the densification of the massif. These methods generally require large soil movements and require a lot of time to run, sometimes requiring areas for disposal and deposits. This work aims to study the use of the constructive method of lightweight embankments, which minimizes the impacts caused by traditional methods, by using expanded polystyrene (EPS) as the main component. The main focus of this work is the case study of the construction of a viaduct on top of embankment in EPS blocks in the municipality of Jundiaí, state of São Paulo, Brazil, comparing the method used with the traditional method of lateral equilibrium berms. During the visit to the work, already completed, presented in the case study, no pathologies were found, showing that the EPS method has potential for use.

1 Introduction

Geotechnical professionals are constantly faced with extraordinarily complex problems in their works, but the challenges are significantly greater when preliminary studies point to soils with predominance of saturated or semi saturated clays with a soft physical aspect. Over the last thirty years, scientific and technological development has led to the emergence of new investigative methods and new materials, with significant contribution to the solution of problems related to the mechanics of these soft soils.

Some conventional construction methods of landfills on soft soils recommend total or partial removal of these soils and the consequent replacement by hard soils. Such replacement favors the stability of the massifs, enabling them to receive the loads foreseen in the project. These methods are reliable and widely used; however, they require considerable time to execute, and at times tit is necessary to overcome long distances

H. Shehata and M. Badr (Eds.): *Advancements in Geotechnical Engineering*, SUCI, pp. 121–135, 2021.
https://doi.org/10.1007/978-3-030-62908-3_9

between the deposit and disposal site, which may, depending on the region, cause great social discomfort, conflict with environmental laws and high cost.

In this study on methods to build embankments in soft soil, Almeida and Marques (2014) describe a method of recent origin, not yet widely used in Brazil, but which has produced satisfactory results in some European countries, Japan and USA. This new construction method recommends the use of low-density materials, which directly influence the weight of the system. The use of such materials contributes to reduce earth movement and locomotion, besides minimizing the stabilization time of embankments with consequent reduction of the total length of the works.

The main geotechnical research tests required to get to know the soft soil intrinsically, with emphasis on the vane test, are an integral part of this work. Notions of horizontal settlements and displacements, types of settlements, stability and monitoring of soft soil embankments have been addressed. A review of the literature also presented some lightweight inert materials and their key characteristics, with emphasis on expanded polystyrene (EPS), stressing the study of its physical-chemical characteristics, the standards governing its manufacture and use, testing and its applicability to lightweight embankments.

This article includes a case study of two embankments of different designs in one single viaduct. On one side, there is an embankment built with equilibrium berms. The other side, due to design needs, was designed as a lightweight Expanded Polystyrene (EPS) embankment, which is the object of a detailed review in this article.

2 Objective

The purpose of this work was to review the feasibility of the construction method of lightweight embankments applied to embankments on top of soft soils using blocks of expanded polystyrene – EPS.

3 Bibliographic Review

3.1 Soft Soils

The term 'soft soil' is closely related to the consistency of certain soils with predominance of clays, often permeated with organic matter almost always in decomposition. These soils are in their natural saturated or semi saturated state. With a sedimentary origin, they have low shear strength and high compressibility, which implies a high capacity of deformation under application of loads (Marangon 2009). Also, according to Marangon (2009), the values determined by the Standard Penetration Test (N-SPT) of soft soils never exceed 5.

3.2 Geotechnical Investigations

According to Almeida and Marques (2014), before starting a geotechnical work intended for construction of embankments on soft soils, it is recommendable to make a survey of the site via geological and pedological maps. The initial studies must be complemented with the aid of aerophotogrammetry and a search in databases of investigations already performed. Subsequently, it is advisable to carry out a preliminary investigation to survey the profile of the terrain by means of percussion soundings or geophysical methods. In the last phase of the geotechnical investigation, field and laboratory investigation must be carried out to obtain the geotechnical parameters, whose basic objective is to provide information to select the construction method and also to provide input to calculate stability and settlement of the embankments. Table 1 states the general characteristics of laboratory and field tests, geotechnical parameters that can be obtained and recommendations for each type of test.

Table 1. Tests available vs. parameters obtained

Type of test	Type of soil		Key characteristics that can be determined
	Best applicable	Not applicable	
1 – standardized penetration test (SPT)*	Granular		Qualitative assessment of the state of compactness or consistency. Qualitative comparison of the stratigraphy of the subsoil
2 – cone penetration test (CPT)	Granular		Continuous assessment of compactness and strength of granular soils. Continuous assessment of undrained strength of clayey soils
3 – vane test	Cohesive	Granular	Undrained strength of clayey soils
4 – pressiometric test	Granular		Thrust coefficient at rest; compressibility and shear strength

*No direct interest in the determination of the strength parameters

(Marangon 2009).

Approximately a dozen tests can be mentioned to determine parameters of soft soils. The detailed description of each test does not fit in this work; however, it is understood that tests producing the basic parameters relevant to decisions on design should be presented, even if briefly. A more detailed description of the vane test, which is the main test when dealing with soft soils, will be presented below.

3.3 Vane Test

Among the tests designed to recognize soft soil parameters, the vane test is the most commonly used. The key purpose of this test is to obtain the shear strength of undrained saturated cohesive soft soils (Su), in situ, according to the Brazilian Association of Technical Standards ABNT NBR 10905 (1989). This test can also determine the sensitivity of clays, which is the ratio between the strength of the undisturbed and disturbed clay.

This test basically consists of driving the vane into the soil and measuring the torque required for its shear on the surface circumscribed by the rotation of the vanes. The positioning of the vane in the measuring height can be done by static driving. The penetration must be stopped and the vane must be placed 50 cm above the height to be measured. From this point on, only the vane descends and the test procedure is carried out.

With the vane at the test height, the crank must be turned at a constant speed of 6°/min, with desirable readings after a 2° rotation, so that the torque vs. rotation curve can be plotted (ABNT NBR 10905 1989).

Right after determining the maximum torque, ten complete revolutions are applied to the vane and the test is repeated to determine the Disturbed strength (S_{ur}).

The time lapse between the two tests shall be less than 5 min.

Interpretation of the Vane Test

According to the Brazilian Association of Technical Standards (ABNT 10905 1989), undrained strength S_u in kPa is given by Eq. 1,

$$Su = 0{,}86\,T \big/ \pi D^3 \left(kN \big/ m^2\right) \tag{1}$$

With torque values supplied by the on-site vane test, where:

T = Maximum torque measured in kN.m.
D = Vane diameter in meters.
Π = 3,1416.

The disturbed strength value (S_{ur}) is obtained by Eq. 1, using, however, the torque value in the collapsed condition.

The sensitivity value (S_t) of clay is given by Eq. 2,

$$St = Su \big/ Sur \tag{2}$$

Figure 1 shows the torque vs. rotation graph of the vane for testing in natural and collapsed clay.

3.4 Settlements

In the construction of a road embankment, soil densification occurs over time, both in the execution phase and in the consolidation phase. Consequently, this soil becomes increasingly harder, leading to an increase in the safety coefficient, but also to an increase in settlements. In the construction period, one must be concerned with the stability of the embankment whereas in the operational phase one must be concerned with settlements. In the case of road embankments, the settlements involve maintenance work to correct wavy roads or adjust the difference in height at bridge and viaduct abutments. In this

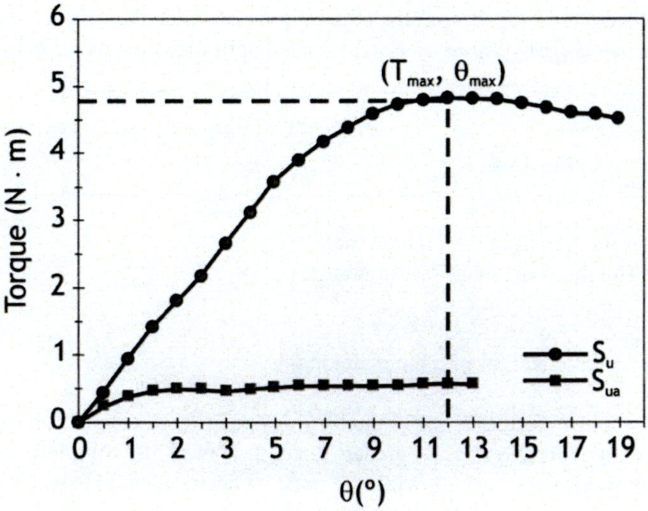

Fig. 1. Torque versus rotation of the vane for testing in natural and collapsed clay. (Neto 2004 apud Almeida and Marques 2014).

context two relevant conditions should be considered, namely the estimation of the final settlements and the estimation of the speed of these settlements (Massad 2010).

According to Das (2011), when a load is applied to a given soil, it is subject to compression and consequent deformation. This phenomenon occurs as a function of the deformation and displacement of soil particles together with expulsion of water or air from hollow spaces. This phenomenon is named settlement.

There are basically three types of settlements: elastic settlement (immediate), settlement by primary densification and settlement by secondary compression.

The total settlement of a foundation can be determined by Eq. 3

$$St = Se + Sc + Ss \quad (3)$$

S_t = Total settlement.
S_e = Elastic settlement.
S_C = Primary settlement.
S_S = Secondary settlement.

It should be noted that, for works on very compressible clays (the key characteristic of soft soils), the settlements by primary densification can be several times larger than the elastic settlement.

3.5 Construction Method with Equilibrium Berms

This construction method uses lateral embankments connected to the main embankment, with lower height, forming the so-called equilibrium berms, whose main objective is to increase the safety factor (F_S) as to rupture. (Almeida and Marques 2014).

Massad (2010) proposes the equation to determine the height of the berms and their respective lengths. Based on Eq. 4 which determines the critical height, we have:

$$\mathrm{H1} - \mathrm{H2} = \frac{\mathrm{S,SS} \cdot C}{F_S \cdot \gamma_{at}} \tag{4}$$

Where:

H_1 = Total height of the embankment.
H_2 = Height of the equilibrium berms.
c = Cohesion.
F_S = Safety factor.
γ_{at} = Specific weight of the embankment.

The safety factor (F_S) is defined by design criteria, considering the relevance of the works. In general terms, F_S greater than 1.5 is used. To complete the dimensioning of the equilibrium berms, the width b2 must be determined as indicated in Fig. 2.

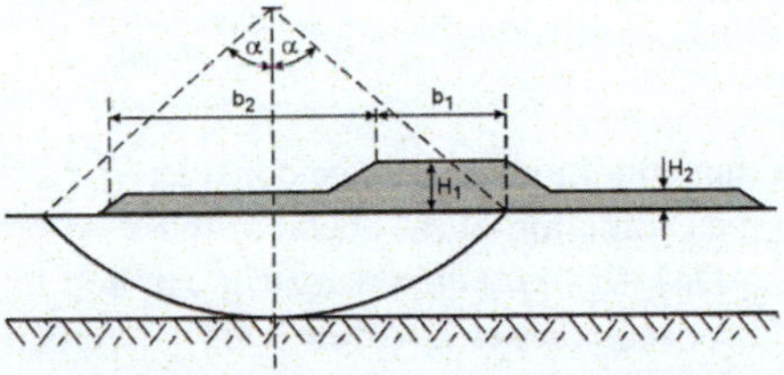

Fig. 2. Equilibrium Berms. (Massad 2010).

3.6 Expanded Polystyrene

What is expanded polyethylene?

Expanded Polystyrene (EPS international abbreviation as defined by ISO 1043 2011), was discovered by chemists Fritz Stasny and Kral Buchholz in Germany in 1949 while working at Basf laboratories (ABIQUIM 2017). Polystyrene polymer (PS), the raw material for EPS (popularly known in Brazil as Isopor®, a registered trademark of Knauf Company) is a stiff, bead-shaped cellular plastic resulting from the polymerization of styrene in water. In the expansion process, pentane gas is used, which is called an expansion agent, and water vapor, causing the beads to increase up to 50 times their original size (ABRAPEX 2008).

Derived from petroleum, a natural product, Fig. 3 shows the production process of PS.

Expanded beads have 98% air and only 2% polystyrene, as shown in Fig. 4 below. The dark region represents the air and the light region represents the polymer, which evidences its characteristics, such as low specific weight, impact damping, easy handling and excellent thermal insulation.

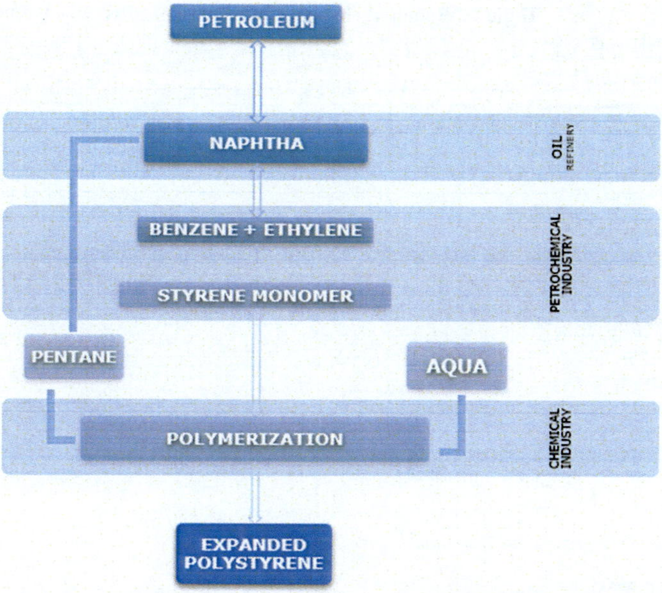

Fig. 3. Production process of EPS. (ABIQUIM 2017).

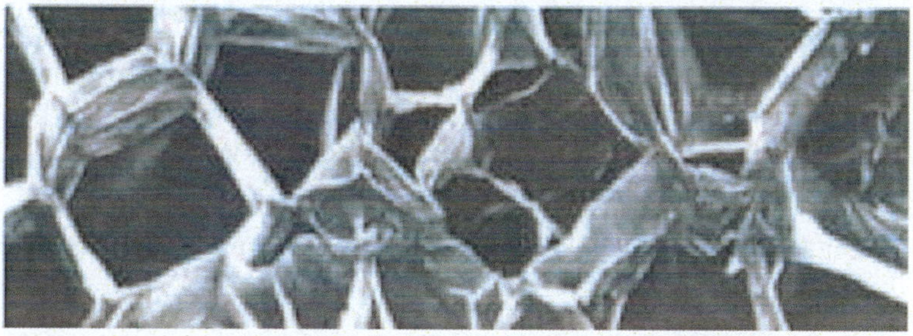

Fig. 4. EPS cell structure under the microscope. (ACEPE 2017).

Classification of expanded polystyrene.

According to the Brazilian Association of Technical Standards (ABNT NBR 11752 2016), expanded polystyrene can be classified into 8 (eight) types, according to its physical properties, specific mass, thermal conductivity, compression stress, flexural strength, water absorption, water vapor permeance, dimensional stability and ignitability. Table 2 presents the characteristics required for expanded polystyrene according to classification by type.

Table 2. Required characteristics for expanded polystyrene

Properties	Test method	Unit	EPS 1	EPS 2	EPS 3	EPS 4	EPS 5	EPS 6	EPS 7	EPS 8
Minimum density	ABNT NBR 11949		10	12	15	18	22	29	38	48
Maximum apparent thermal conductivity (24 °C)	ASTM C 177 or ASTM C 518	W/(m.K)	0.046	0.046	0.040	0.038	0.036	0.034	0.034	0.033
Minimum stress by compression with 10% deformation	ABNT NBR 8082	kPa	35	35	69	90	104	173	276	414
Minimum flexural strength	ASTM C 203	kPa	50	70	173	208	240	345	414	517
Maximum water absorption	ASTM C 272	% in volume	6	4	4	3	3	2	2	2
Maximum permeance of water vapor in 25.4 mm	ASTM E 96/E 96M	ng/Pa.s.m^2	345	287	287	201	201	143	143	143
Dimensional stability	ASTM D 2126	%	2.0	2.0	2.0	2.0	2.0	2.0	2.0	2.0
Ignitability	EM SSO 11925-2	$Fs \leq 150$ mm in 20 s								

(ABNT NBR 11752 2016).

3.7 Embankment using Blocks of Expanded Poystyrene - EPS

EPS stands out since, in comparison to other materials that lend themselves to the same use, it has a significantly lower specific weight. Moreover, its physical characteristics have the advantage of high resistance (70 to 250 kPa) and low compressibility, with modulus of elasticity in the order of 1 to 11 Mpa, varying according to the specific weight. According to Almeida and Marques (2014), the physical variance of EPS, with different densities and consequent resistance to stress should be the object of choice in projects, according to the mobile loads that will be applied in the system. Figure 5 presents a construction in North America executed with blocks of EPS, in order to provide reduced loads on the underlying layers, thus contributing to stabilize the slope.

The sides and top surfaces of embankments made of EPS must include a layer of soil. The lateral containments can be coated with sprayed concrete; the upper surface should get a reinforced concrete slab with thickness dimensioned to distribute the stresses on the EPS, thus avoiding puncturing the light material due to dynamic and punctual loads of vehicle traffic. EPS does not resist the action of organic solvents. Therefore, it is advisable to provide coating with a high-density polyethylene (HDPE) blanket, which is a highly stable polymer and consequently insensitive to harmful liquids. Figure 6 shows the cross-section of an EPS embankment. The design shows that there is a layer of soil (h_{at}) on the top surface of the embankment, which varies in height according to the use

Fig. 5. Use of blocks of EPS to stabilize slopes. (Geofoam Applications and Technical Data 2016).

of the embankment, e.g. permissible load limit on the road and traffic density. At the site where an EPS embankment is to be installed, the level of the water table must be checked at a time of high rainfall and possible flooding. The EPS should be seated at a positive height in relation to these levels, or a study of the water-flood thrust should be made to ensure stability of the system (Almeida and Marques 2014).

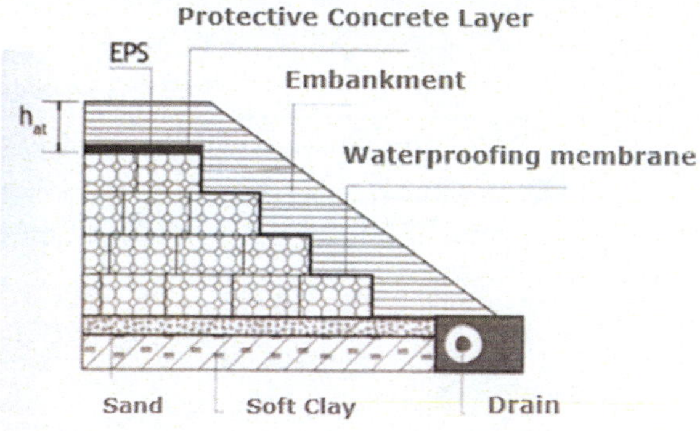

Fig. 6. Cross-section of an embankment built with EPS blocks. (Almeida and Marques 2014)

3.8 Case Study

Google Earth made it possible to view the place in the year of 2010, as seen in Fig. 7. This photo was taken after the execution of the viaduct that leads to Caxambu neighborhood, on Eng. Constâncio Contra (SP 360) Road, km 67.

Characteristics of the Place

According to the Fundações e Obras Geotécnicas Magazine (2014), the interchange was

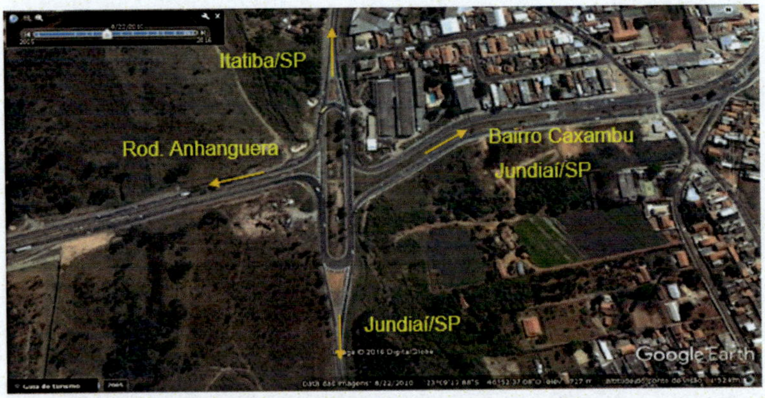

Fig. 7. Photo of the place before the embankment with EPS blocks was made. Year: 2010. (Google Earth 2017)

built because of the requirement to build a road complex to help with the large flow of vehicles in the Caxambu interchange, worsened on weekends, in the region of Itatiba, São Paulo, Brazil.

As presented at the 1st International Workshop on Trends of EPS as a Constructive Solution (2015), the region where the viaduct was built has successive layers of soils composed of soft organic silty-sandy clay. Figure 8 shows the geotechnical study of the site, according to the Standard Penetration Test (SPT).

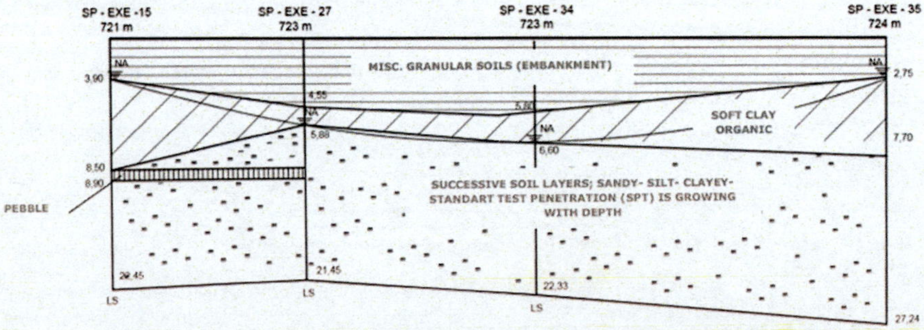

Fig. 8. Profile of SPT percussion sounding. (Costa Junior and Vasquez 2015).

Method of Execution Adopted

A review of the geotechnical investigations of the region including the location where the viaduct was to be built, which was fully composed of soft soils, indicated the use of expanded polystyrene blocks (lightweight embankment method) together with geosynthetic materials. Figure 9 depicts the model of the geoproject proposed.

Figure 10 shows the longitudinal section, placement of drains and indications.

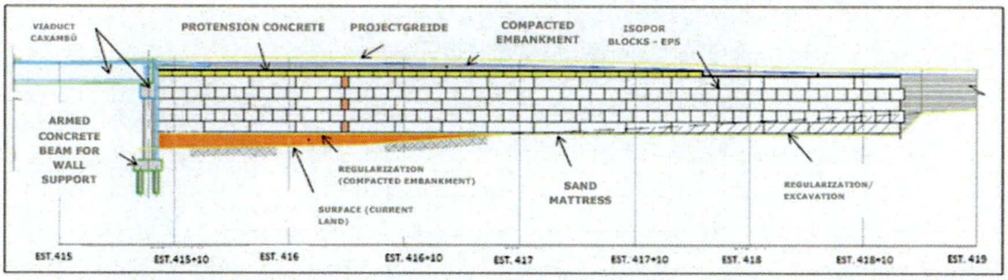

Fig. 9. Geoproject proposed for a lightweight embankment with EPS blocks. (Costa Junior and Vasquez 2015)

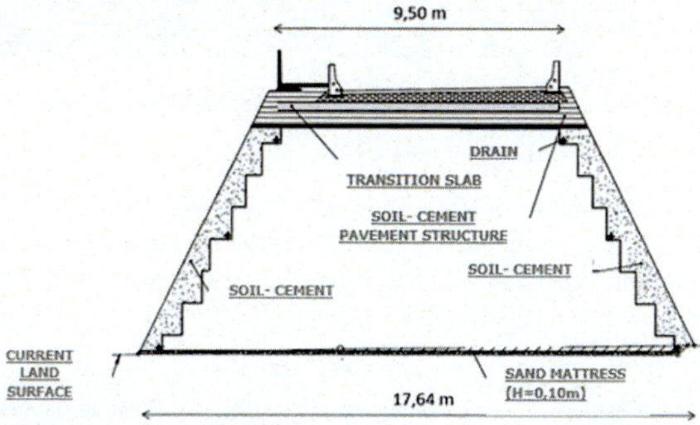

Fig. 10. Longitudinal section along the access piling axis. (Costa Júnior and Vasquez 2015)

Blankets of high-density polyethylene were installed to protect the blocks of expanded polystyrene from attack by organic solvents and others. Figure 11 provides the specification for the blanket and installation of drains; this item is critical to drain water from infiltrations.

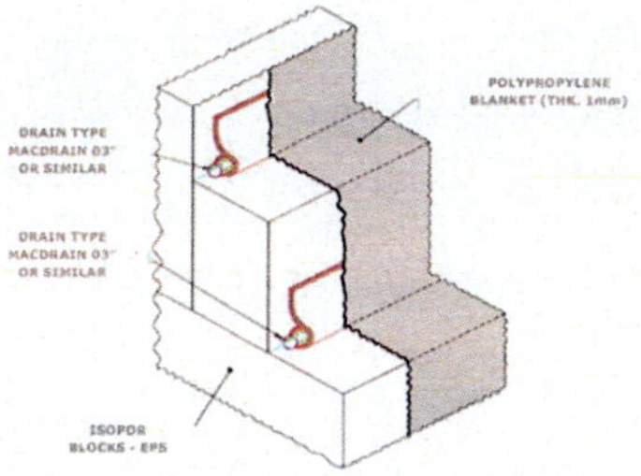

Fig. 11. Details of the drain and HDPE blanket. (Costa Júnior and Vasquez 2015)

Figure 12 shows the placement of the subsequent layers, which strictly follow the modulation design and use of the 23 kg/m^3 blocks, totaling 8 overlapping layers on a curve, according to the projected layout of the road.

Fig. 12. Placement of the subsequent layers of EPS blocks. (Costa Junior and Vasquez 2015).

To protect the top of the blocks and beneath the 1-mm PEAD blanket, a reinforced concrete slab (10 cm) was built with two layers of 10 × 10 cm welded mesh, TELCON type or equivalent, with CA 60 steel bars, 4.2-mm diameter and C30 concrete, as shown in Fig. 13.

Fig. 13. Execution of the reinforced concrete slab. (Costa Júnior and Vasquez 2015).

On-site Visit An on-site visit made on April 9, 2017 demonstrated that the integrity of the embankment and viaduct was qualitative. Figure 14 shows the visit to the embankment executed via the method of equilibrium berm. It is possible to see the area taken by the lateral berms.

Comparison Between the Method of Equilibrium Berms and the Method of Lightweight Embankment At the end of this paper, the authors organized Table 3

Fig. 14. View of the embankment executed with the method of equilibrium berms. (Authors 2017)

to compare the construction methods of lightweight embankment and the method of equilibrium berms.

Table 3. Comparison of the methods described

Equilibrium berms	Lightweight embankment
Large area taken due to the berms	Small area taken, no berms
Major movements of soil	Little movement of soil
Large movement of machines and equipment	Little movement of machines and equipment
Deposit needed	Regional EPs manufacturers required
Disposal area required in case of replacement	The method uses disposal area only for regularization of the base
Heavy weight of the massif itself	The massif itself has very low weight
First step is to build the embankment and then the works	Execution together with the works and embankment
Large primary settlements	Negligible settlements
Large viaduct/embankment differential settlements	Small viaduct/embankment differential settlements
High risk of instability right after the execution	No risk of instability right after the execution
No need for constant topography	Constant topography required
Unfeasible execution on rainy days	Can be executed on rainy days
Concerns with attacks by petroleum byproducts are not applicable	Requires protection because it is sensitive to attacks of petroleum solvents

(Authors 2017)

4 Conclusion

After a qualitative analysis of the case study, on-site visit and together with a review of the literature, the authors conclude:

No cracks were visually observed in the longitudinal and in the cross-sectional direction of the paving on the lightweight embankment.

A close look at the abutment between the viaduct and the lightweight embankment showed no secondary differential settlement visually.

No holes in the paving were detected that might display fines pumping.

No wavy runways.

The drains of the lightweight embankment (all of them internal) were operating efficiently. This was corroborated by the absence of significant signs of water running down the sides of the lightweight embankment.

On the south face of the lightweight embankment, cracks of 2 mm to 5 mm were detected in the projected concrete. Such cracks are being monitored. The monitoring with glass sheets indicated rupture, demonstrating signs of increase and/or expansion of the cracks.

It was not possible to check the depth of the cracks.

The heads of the viaduct seemed preserved.

The traffic flowed normally, with no restrictions on the amount and type of vehicles.

No road signs were seen to restrict the use of the viaduct and embankment.

No pathologies were found that might call the attention of the authors of this work and would require warnings about any condemnation of the works made.

The use of Expanded Polystyrene (EPS) in road embankments on soft soil is technically feasible. The so-called Lightweight Embankment has become a constructive option for the geotechnical sector.

Acknowledgements. We wish to thank the GeoMEast 2019 International Congress and Exhibition - for the possibility of recognition of this article at international level.

We thank our masters Roberto Kassouf and David Carvalho from whom we have learned Engineering in its essence and for the guidance they have given us along this journey.

We thank all UniMetrocamp Wyden faculty members who contributed to our learning day after day and to whom we are immensely grateful for the legacy awarded.

We thank Nicolas Eduardo Vasquez and Milton Paulino Costa Junior (engineers) who, through the 1st International Workshop on trends of EPS as a constructive solution, have helped develop this work, with the presentation of the Odebrecht Infrastructure Case.

References

ABIQUIM: Poliestireno Expansível. Comissão Setorial. http://www.epsbrasil.eco.br/eps/index.html#caracteristicas. Accessed 15 Feb 2017

ACEPE: Associação Industrial do Poliestireno Expandido de Portugal. http://acepe.pt/index.php/2013-04-03-16-41-27/documentacao-livros-manuais-e-estudos. Accessed 7 Mar 2017

de Almeida, M.S.S.: Aterros sobre solos moles da concepção à avaliação do desempenho. Editora UFRJ, Rio de Janeiro (1996) de Almeida, M.S.S., Marques, M.E.S.: Aterros sobre solos moles projeto e desempenho. Editora Oficina de Textos, São Paulo (2014)

Associação Brasileira de Normas Técnicas – ABNT: NBR 10905. Solo – Ensaio de palheta in situ – Método de ensaio. Rio de Janeiro (1989)

Associação Brasileira de Normas Técnicas – ABNT: NBR 11752. Materiais celulares de poliestireno para isolamento térmico na construção civil e refrigeração industrial. Rio de Janeiro (2016)

de Bueno, B.S., Neto, J.O.A.: Ensaio de absorção de água em blocos de EPS para uso geotécnico. Instituto de Pesquisas Tecnológicas do Estado de São Paulo – IPT, São Paulo (2009)

Costa Junior, M.P., Vasquez, N.E.: 1° Workshop internacional sobre tendências do EPS como solução construtiva. 1, 2015, São Paulo. ABIQUIM: Apresentação Case Odebrecht Infraestrutura, São Paulo (2015)

Crespo Neto, F.N.: Efeito da velocidade de rotação na tensão cisalhante obtida em ensaio de palheta. Dissertation (Master) - COPPE/UFRJ, Rio de Janeiro (2004)

Das, B.M.: Fundamentos de engenharia geotécnica, 7th edn. Cengage Learning, Norte Americana (2011)

EPS Industry Alliance Geofoam Civil Engineering: Geofoam Applications & Tecnhical Data. http://www.geofoam.com. Accessed 25 Sep 2016

Google Earth. https://earth.google.com/web/. Accessed 8 Feb 2017

International Organization for Standardization – ISO 1043: Plastics – Symbols and abbreviated terms – Part 1: Basic polymers and their special characteristics, London (2011)

Marangon, M.: Resistência ao cisalhamento dos solos. UFJF, Juiz de Fora (2009). http://www.ufjf.br/nugeo/files/2009/11/ms2underlineunide05-p1.pdf. Accessed 6 Feb 2017

Massad, F.: Obras de terra. Curso básico de geotecnia. 2nd edn. Oficina de Textos, São Paulo (2010)

Scatena, P., Scatena, V.: Poliestireno Expandido em Aterros dobre Solos Moles. Monography (Graduation) – Devry Metrocamp, Campinas, Brazil (2017)

Future of Underground Techniques in Deep Mines of Different Content Without Catastrophic Risks

Olga Hachay[1] and Oleg Khachay[2(✉)]

[1] Institute of Geophysics UB RAS, named after Yu.P.Bulashevitch, Yekaterinburg, Russia
[2] Ural Federal University, named after first President of Russia B.N.Eltsin, Yekaterinburg, Russia
olgakhachay@yandex.ru

Abstract. Self-organization is not a universal property of matter, it exists under certain internal and external conditions and this is not associated with a special class of substances. The study of the morphology and dynamics of migration of anomalous zones associated with increased stresses is of particular importance in the development of deep deposits, complicated by dynamic phenomena in the form of rock impacts. An important tool for this study is geophysical exploration. To describe the geological environment in the form of an array of rocks with its natural and technogenic heterogeneity, one should use its more adequate description, which is a discrete model of the medium in the form of a heterogeneous block medium with embedded heterogeneities of a lower rank than the block size. We have carried out an analysis of the morphology of the structural features of disintegration zones before a strong dynamic phenomenon. The constructed theory demonstrated how complicated the process of complexion methods using an electromagnetic and seismic field is to study the response of a medium with a hierarchical structure. This problem is inextricably linked with the formulation and solution of the inverse problem for the propagation of electromagnetic and seismic fields in such complex environments. Using the theory of solving the inverse problem, one can trace the migration of zones of abnormal stresses, their increase or decrease due to cyclic explosive influences during mining of the array, and also assess the possible risk of high-energy dynamic phenomena in the array. The most important conclusion consists on that fact: the monitoring system must be tuned to search the rock massif as a hierarchical structure. The catastrophic events are linked with the dynamics of the hierarchical inclusions state. For today we have sufficient knowledge to use these systems with mathematical software of predicting catastrophic events.

Keywords: Hierarchical environment · Electromagnetic · Acoustic field · Iterative algorithm · Integral and integro-differential equations · Direct problem

H. Shehata and M. Badr (Eds.): *Advancements in Geotechnical Engineering*, SUCI, pp. 136–143, 2021.
https://doi.org/10.1007/978-3-030-62908-3_10

1 Introduction

The formation of structures during irreversible processes is associated with a qualitative leap when threshold (critical) parameters are reached. Self-organization is a supercritical phenomenon when the parameters of the system exceed their critical values. When a system deviates strongly from an equilibrium state, its variables satisfy non-linear equations. Nonlinearity is an important and general feature of processes that occur far from equilibrium. Moreover, the supercritical return of entropy is possible only if there is an unusual, special internal structure of the system (Ebeling 1979). This means that self-organization is not a universal property of matter, it exists under certain internal and external conditions and this is not associated with a special class of substances. So, there are two classes of irreversible processes: 1. destruction of the structure near the equilibrium position, this is a universal property of systems under arbitrary conditions; 2. the appearance of structures far from the equilibrium position under the conditions that the system is open and has nonlinear internal dynamics, and its external parameters have supercritical values. I. Prigogine called them dissipative structures (Glensdorf et al. 1973). The study of the morphology and dynamics of migration of these zones (named as disintegration zones) is of particular importance in the development of deep deposits, complicated by dynamic phenomena in the form of mountain impacts. An important tool for this study is geophysical exploration. As shown in (Sadovskiy et al. 1987), to describe the geological environment in the form of an array of rocks with its natural and technogenic heterogeneity, one should use its more adequate description, which is a discrete model of the medium in the form of a piecewise inhomogeneous block medium with embedded heterogeneities of lower rank than the block size. This nesting can be traced several times, i.e. changing the scale of the study; we see that heterogeneities of a lower rank now appear in the form of blocks for heterogeneities of the next rank. A simple averaging of the measured geophysical parameters can lead to distorted ideas about the structure of the medium and its evolution (Sadovskiy et al. 1987). One of the fundamental mining problems, which are traditionally referred to the problems of geomechanics, is the development of theoretical and experimental methods for studying the structure and condition of rock masses in order to predict and prevent catastrophic phenomena during mining. This problem is compounded by the fact that the rock massif is under direct or indirect technogenic impact, which leads to significant unsteadiness of both the structure and the state of the mass (Hachay et al. 1987). When conducting mining operations in highly stressed rock massifs, technogenic seismicity is manifested, the forecasting and prevention issues of which receive a lot of attention in all countries with a developed mining industry. An important role here belongs to the short-term forecast; the methodology for selecting criteria for it is still a problem, both in mining and in seismology (Kozyrev et al. 2001). In the framework of the Institute of Mining of the Siberian Branch of Russian Academy of sciences, important results have been achieved on the study of the state of a rock massif in the framework of nonlinear geomechanics (Kurlenja et al. 1999) using geophysical methods with a resolution to detect the nucleation and decay of self-organizing structures (Hachay 2006). In (Hachay 2006), studies were conducted aimed at developing criteria for the spatial-temporal complex active and passive seismic and electromagnetic monitoring to prevent destructive dynamic phenomena based on six-year seismological monitoring data carried out by the mountain impact service at the Tashtagol underground mine and the experience gained use of the system of induction electromagnetic spatial-temporal monitoring developed at the Institute of geophysics of Ural Branch of Russian Academy of sciences for arrays of various material composition before and after mass explosions.

2 Theory, Research Methods

We analyzed the morphology of the structural features of disintegration zones before a strong dynamic phenomenon. During the next cycle of electromagnetic observations at the Tashtagol mine in August 2007 On August 9, a mountain impact occurred with an energy of log E = 6.9 in the pillar, located in the orth 3 alignment at a level of 16 m below the soil of horizon −280, Fig. 1.

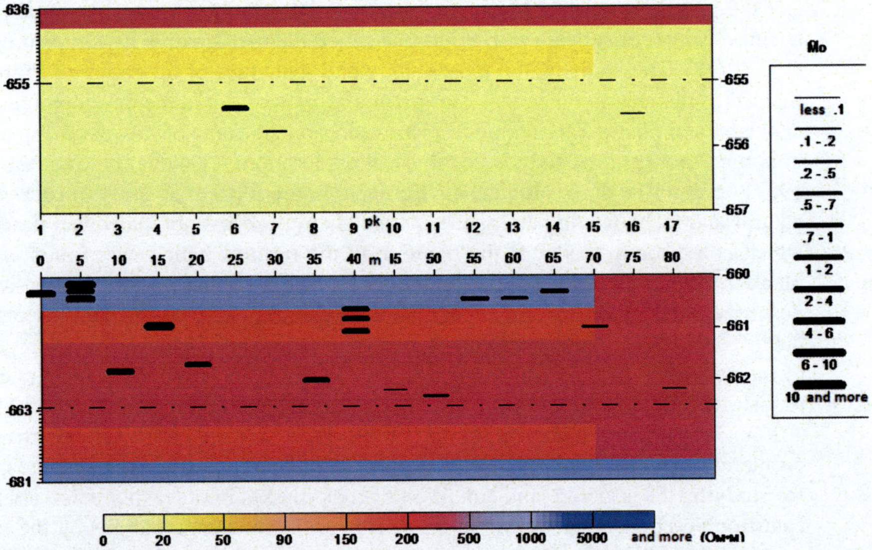

Fig. 1. Geoelectrical section of a block-layered section of rock massif with inclusions of the second rank in a rock burst mine Tashtagol (the upper half of the section corresponds to the roof structure, the lower – to the floor) Morphology of anomaly stressed objects.

Legend: = = M0 × L0 × 103, M0 is the coefficient by which the moment of the electric current line is multiplied, equivalent in field to the influence of the zone of geoelectric heterogeneity, and which is proportional to the ratio of the difference in conductivities in the enclosing medium and in inclusion to the conductivity in the enclosing medium, L0 - the length of the current line, the resistance of the enclosing section is given in ohm.m. The vertical values are in m (absolute marks), the horizontal is the length of the output in pickets (pc) and meters.

Three days before a mountain shock, in the orth 3 (Fig. 1, 2), in the geoelectric sections of the soil, sub vertical discrete structures are found in which disintegration zones are combined. These structures appeared in the resonance mode at different frequencies and only at one frequency for each of the unit vectors. We discovered the same phenomenon earlier in one day at Estyuninsky mine and SUBR mine 15 (Hachay et al. 2007a). The appearance of these structures of sub vertical morphology is a precursor of a strong dynamic phenomenon; however, to determine the place and magnitude of an event, it is necessary to have information on the state of the unit arrays and their belonging to the corresponding ranks on the stability of the array, as was done in (Hachay 2007).

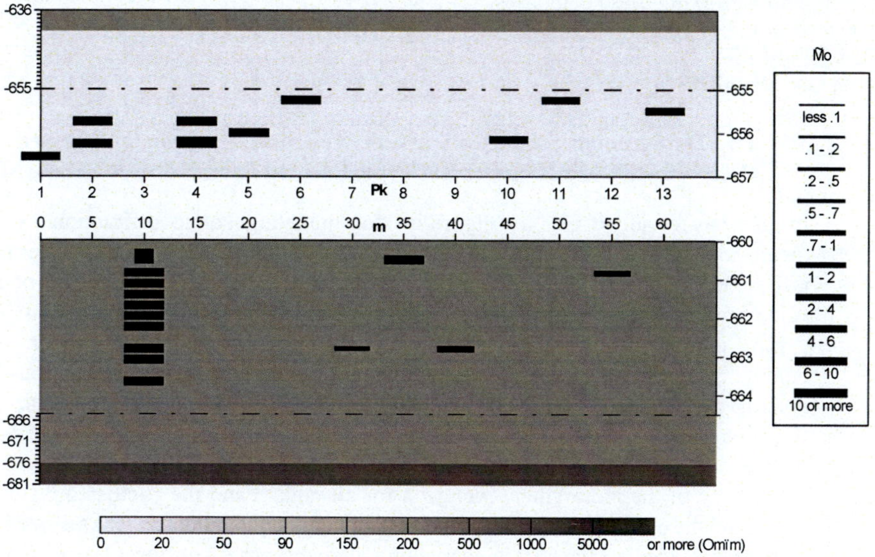

Fig. 2. Geoelectric section along the profile of ort3, horizon-210, North-Western section, August 7, 2007, frequency 5.08 kHz. The legend is the same as in Fig. 1.

At present, theoretical results on modeling the electromagnetic and seismic fields in a layered medium with hierarchical inclusions are in demand. Modeling algorithms were constructed in the electromagnetic case for 3D heterogeneity, in the seismic case for 2D heterogeneity (Hachay et al. 2015, 2016a). It is shown that with an increase in the degree of hierarchy of the medium, the degree of spatial non-linearity of the distribution of components of the seismic and electromagnetic fields increases, which corresponds to the detailed monitoring experiments in shock hazardous mines of the Tashtagolsky mine and SUBR. The constructed theory demonstrated how complicated the process of complexion methods using an electromagnetic and seismic field is to study the response of a medium with a hierarchical structure. This problem is linked with the formulation and solution of the inverse problem for the propagation of electromagnetic and seismic fields in such complex media. In (Hachay et al. 2016b, 2017), the problem of constructing an algorithm for solving the inverse problem using the equation of the theoretical inverse problem for the 2D Helmholtz equation was considered. Explicit equations of the theoretical inverse problem are written out for the cases of scattering of an electromagnetic field (E and H polarization) and scattering of a linearly polarized elastic wave in a layered conducting and elastic medium with a hierarchical conducting or elastic inclusion, which are the basis for determining the contours of misaligned inclusions l-rank of hierarchical structure. Obviously, when solving the inverse problem, monitoring systems configured to study the hierarchical structure of the environment should be used as the initial monitoring data. On the other hand, the more complex the medium, the each wave field brings its own information about its internal structure; therefore, the interpretation of the seismic and electromagnetic fields must be carried out separately, without mixing these databases.

3 Results

3.1 Two-Dimensional Sound Diffraction Modeling Abnormally Stressed Heterogeneity of a Hierarchical Type, Located in an N-Layer Elastic Medium

In (Hachay et al. 2011), an algorithm for modeling sound diffraction by a two-dimensional elastic hierarchical inclusion located in the J-th layer of an N-layer medium is described $G_{Sp,j}(M, M_0)$ – the function of the source of the scismic field, the boundary-value problem for which was formulated in (Hachay et al. 2011); $k_{1,ji} = \omega^2(\sigma_{ij}/\lambda_{ij})$ – wave number for a longitudinal wave, in the above expression, the index ji means belonging to the properties of the medium inside the heterogeneity, ja is outside the heterogeneity, λ is the Lame constant; σ is the density of the medium; ω is the circular frequency; $\vec{u} = grad\varphi$ - vector of displacements; φ^0 – is the potential of a normal seismic field in a layered medium in the absence of heterogeneity: $\varphi^0_{ji} = \varphi^0_{ja}$. We assume that the density of the hierarchical inclusion for all ranks l and the enclosing layer are the same, and the elastic parameters of the hierarchical inclusion for all ranks differ from the elastic parameters of the enclosing medium, then the system of Eqs. (3) can be rewritten in the form:

$$\frac{(k^2_{1jil} - k^2_{1j})}{2\pi} \iint\limits_{SCl} \varphi_l(M) G_{Sp,j}(M, M^0) d\tau_M + \varphi^0_{l-1}(M^0) = \varphi_l(M^0),\ M^0 \in S_{Cl},$$

$$\frac{\sigma_{jil}(k^2_{1jil} - k^2_{1j})}{\sigma(M^0)2\pi} \iint\limits_{SCl} \varphi_l(M) G_{Sp,j}(M, M^0) d\tau_M + \varphi^0_{l-1}(M^0) = \varphi_l(M^0),\ M^0 \notin S_{Cl}. \qquad (1)$$

3.2 Elastic Shear Wave Diffraction Modeling on an Abnormally Stressed Heterogeneity of a Hierarchical Type, Located in an N-Layer Elastic Medium

Similarly to (1), the same process is written for modeling the propagation of an elastic transverse wave in an N-layer medium with a two-dimensional hierarchical structure of an arbitrary section morphology using the integral relations written in (Hachay et al. 2015).

$$\frac{(k_{2jil}^2 - k_{2j}^2)}{2\pi} \iint\limits_{SCl} u_{xl}(M) G_{Ss,j}(M, M^0) d\tau_M + \frac{\mu_{ja}}{\mu_{jil}} u_{x(l-1)}^0(M^0)$$
$$+ \frac{(\mu_{ja} - \mu_{jil})}{\mu_{jil} 2\pi} \oint\limits_{Cl} u_{xl}(M) \frac{\partial G_{Ss,j}}{\partial n} dc = u_{xl}(M^0),\ M^0 \in S_{Cl},$$
$$\frac{\mu_{jil}(k_{2jil}^2 - k_{2j}^2)}{\mu(M^0)\, 2\pi} \iint\limits_{SCl} u_{xl}(M) G_{Ss,j}(M, M^0) d\tau_M + u_{x(l-1)}^0(M^0)$$
$$+ \frac{(\mu_{ja} - \mu_{jil})}{\mu(M^0)\, 2\pi} \oint\limits_{Cl} u_{xl}(M) \frac{\partial G_{Ss,j}}{\partial n} dc = u_{xl}(M^0),\ M^0 \in S_{Cl}. \quad (2)$$

$G_{Ss,j}(M, M^0)$ – the source function of the seismic field of the problem under consideration, it coincides with the Green function written in (Hachay et al. 2015) for the corresponding problem;

$k_{2jil}^2 = \omega^2(\sigma_{jil}/\mu_{jil})$, $\mu_{jil} \neq \mu_{ja}$ – shear wave number, $\sigma_{jil} = \sigma_{ja}$; μ – Lame constant; u_{xl} – component of the displacement vector; $l = 1 \ldots L$ – hierarchical level number; u_{xl}^0- component of the displacement vector of the seismic field in a layered medium in the absence of heterogeneity of the previous rank, if $l = 2 \ldots L$, $u_{xl}^0 = u_{x(l-1)}$, if $l = 1$, $u_{xl}^0 = u_x^0$, which coincides with the corresponding expression for the normal field in the work (Hachay et al. 2015). It should be noted that the structure of Eqs. (4) coincides with the general case when the hierarchical heterogeneity has not only elastic parameters different from the parameters of the enclosing medium, but also density parameters at all ranks differ from the density parameters of the enclosing layer. The difference between these problems lies only in the values of the wave number. Thus, the response of the medium associated with the longitudinal wave is more sensitive to the region of elastic heterogeneities in the array. This should be taken into account when assessing the status of a complex organized geological environment. In (Hachay et al. 2017), the problem of constructing an algorithm for solving the inverse problem using the equation of the theoretical inverse problem for the 2-D Helmholtz equation was considered. An explicit equation of the theoretical inverse problem is obtained for the cases of scattering of a linearly polarized elastic wave in a layered elastic medium with a hierarchical elastic inclusion, the density of which for all ranks is equal to the density of the containing layer. An iterative algorithm for determining the contours of misaligned inclusions of the k-th rank in a hierarchical structure is constructed with the sequential use of the solution of the direct problem of calculating the elastic field of k-1 rank. With an increase in the degree of hierarchy of the structure of the medium, the degree of spatial non-linearity of the distribution of the components of the seismic field increases, which implies the exclusion of linearization methods for creating interpretation methods. This problem is linked with the solution of the inverse problem for the propagation of the seismic field in such complex environments using explicit equations of the theoretical inverse problem. For the first time, an equation was written for determining the surface of an abnormally stressed inclusion in a hierarchical layered-block medium according to acoustic monitoring data. In practice, using this algorithm, according to acoustic monitoring data, we can localize the region of a possible source of a rock shock or an earthquake and estimate the degree of anomalous elastic stresses.

4 Discussion of Results and Conclusions

Comparing expressions (1) and (2) we can draw the following conclusions. When constructing a complex seismic-gravity model without taking into account the anomalous influence of the stress-strain state inside the inclusion, the analysis of the anomalous acoustic effect using the data on the propagation of a longitudinal wave shows that it is also more sensitive to the form of inclusion, compared with the acoustic effect on the propagation of a transverse wave. However, it follows from these expressions that the influence of the elastic parameters in the host medium in the seismic model cannot be neglected, and they, when interpreted, affect the values of the desired anomalous densities. If these values are used in constructing the gravitational density model, then these density values will not reflect the material composition of the analyzed medium. When constructing an anomalously stressed geomechanical model without taking into account the anomalous effect of density heterogeneities within the inclusion, an analysis of the anomalous acoustic effect using data on the propagation of a shear wave shows that it is also more sensitive to the form of inclusion, compared with the acoustic effect on the propagation of a longitudinal wave. However, it follows from these expressions that the influence of the density parameters in the host medium in the seismic model cannot be neglected, and when interpreted, they affect the values of the desired anomalous elastic parameters that cause the anomalous stress state. If these values are used in the construction of the geomechanical model, then these values of the elastic parameters will not reflect the stress state of the analyzed medium. It is shown that with an increase in the degree of hierarchy of the medium, the degree of spatial non-linearity of the distribution of components of the seismic and electromagnetic fields increases, which corresponds to the detailed monitoring experiments in shock-hazardous mines of the Tashtagolsky mine and SUBR. The constructed theory demonstrated how complicated the process of complexion methods using an electromagnetic and seismic field to study the response of a medium with a hierarchical structure. This problem is linked with the formulation and solution of the inverse problem for the propagation of electromagnetic and seismic fields in such complex media. In (Hachay et al. 2017), the problem of constructing an algorithm for solving the inverse problem using the equation of the theoretical inverse problem for the 2D Helmholtz equation was considered. Using the theory of solving the inverse problem, one can trace the migration of zones of abnormal stresses, their increase or decrease due to cyclic explosive influences during mining of the array, and also assess the possible risk of high-energy dynamic phenomena in the array.

In the book Fedorchenko et al. (1981) had been suggested a new theoretical approach to identify the type of medium stability using the features of the dispersion equations for penetrating plane waves through a homogeneous medium. Now we are elaborating an algorithm for the case of hierarchic medium.

References

Ebeling, V.: The formation of structures during irreversible processes, Mir, Moscow (1979)

Glensdorf, P., et al.: Thermodynamic theory of structure, stability and fluctuations, Mir, Moscow (1973)

Hachay, O.A.: The problem of studying the transition process of redistribution of the stressed and phase states of the array between strong technogenic effects. Min. Inf. Anal. Bull. **5**, 109–115 (2006)

Hachay, O.A.: The study of the instability development in a rock massif using the method of active electromagnetic monitoring. Phys. Earth **4**, 65–70 (2007)

Hachay, O.A., et al.: Three-dimensional electromagnetic monitoring of the state of a rock massif. Phys. Earth **2**, 85–92 (1987)

Hachay, O.A., et al.: The results of geophysical and geological research at the Estyuninsky mine. In: Proceedings of the Fourth scientific readings in memory of Yu. P. Bulashevitch "Deep structure, geodynamics, thermal field of the Earth, interpretation of geophysical fields." Yekaterinburg, IGF UB RAS, 02–06 July 2007 Ekaterinburg, IGF UB RAS, 197–199 (2007a)

Hachay, O.A., et al.: On the integration of seismic and electromagnetic active methods for mapping and monitoring the state of two-dimensional heterogeneities in an N-layer medium. Bull. SUSU. Ser. Comput. Technol. Manage. Electr. **2**, 49–56 (2011)

Hachay, O.A., et al.: Modeling of electromagnetic and seismic fields in hierarchically heterogeneous media. Bull. SUSU, Ser. Computat. Math. Comput. Sci. **2**(2), 48–55 (2013)

Hachay, O.A., et al.: New methods of Geoinformatics of monitoring wave fields in hierarchical environments. Geoinformatics **3**, 45–51 (2015)

Hachay, O.A., et al.: New methods of geoinformatic for the integration of seismic and gravitational fields in hierarchical environments. Geoinformatics **3**, 25–29 (2016a)

Hachay, O.A., et al.: Determination of the surface of anomalously stressed inclusions in a hierarchical layered-block medium according to acoustic monitoring. Min. Inf. Anal. Bull. **4**, 354–356 (2016b)

Hachay, O.A., et al.: On the inverse problem of active electromagnetic and acoustic monitoring of a hierarchical geological environment. Geophys. Res. **18**(4), 71–84 (2017). https://doi.org/10.21455/gr2017/4-6

Fedorchenko, A.M., et al.: Absolute and convective instability in plasma and solids. "SCIENCE" Main edition of physical and mathematical literature, Moscow (1981)

Kozyrev, A.A., et al.: Features of the forecast and prevention of powerful dynamic phenomena in natural-technical systems. In: Proceedings of the international conference "Geodynamics and stress state bowels of the Earth ". Novosibirsk, IGD SB RAS, Russia, 2–4 October 2001. Novosibirsk: IGD SB RAS, Russia, 326–334 (2001)

Kurlenja, M.V., et al.: Current problems of nonlinear geomechanics. Proceedings of the international conference "Geodynamics and stress state of the bowels of the Earth". Novosibirsk IGD SB RAS, 6–10 July 1999 - Novosibirsk IGD SB RAS, 5–20 (1999)

Sadovskiy, M., et al.: Deformation of the geophysical environment and the seismic process, Nauka, Moscow (1987)

Utilization of Silica Fume and Slag to Improve the Geotechnical Properties of Expansive Soil

Ahmed Hussien Zaki(✉), Samir Gad, and Mahmoud AbuBakr Alsedik

Soil Mechanics and Foundation, Al-Azhar University, Cairo, Egypt
ahmed_zaki@azhar.edu.eg

Abstract. Expansive soils are recognized as a problematic soil that imposes several challenges for civil engineers. Such soil undergo significant volume change in case of water penetrates into them, and they shrink as they lose moisture. Lightly-loaded engineering structures such as pavements, single story buildings, railways and walkways may experience severe damages when they are founded on such soils. Determination of expansive soils and quantifying their swelling potential and pressure caused by their expansion are essential in the application of geotechnical engineering practice. Therefore it is necessary to overcome problems associated with the swelling potential through introducing some other materials as an improvement aids.

The present study presents an experimental investigation of the swelling behavior of an expansive soil taken from Al-Andalos area at New Cairo, Egypt. Atterberg limits, Grain size distribution, Specific gravity, Optimum water content, maximum dry density, Swelling pressure, Free swell index (FSI) and swell potential of the soil were assessed. Based on the test results, soil sample classified as a highly clay percentage which leads to activity. Silica Fume and Slag were added to soil sample with ratios of 1%, 3%, 5%, 6%, 7% and 10% by dry weight of soil sample as an improvement and their effect on the geotechnical properties of tested soil were investigated.

The obtained test results were analyzed to establish the optimum percentage that can be added to improve the geotechnical properties of expansive soil in terms of swelling potential and unconfined compressive strength. Based on the obtained results, the treatment of the expansive soil with 5% of both Silica fume and Slag revealed better improvement with moderate swelling potential. Otherwise, the unconfined compressive strength should uncertainty trend depending on the shear parameter characteristics of both Silica fume and Slag.

Keywords: Expansive soil · Silica fume · Slag · Swelling potential · Free swell index · Swelling pressure · Unconfined compressive · Standard proctor test

1 Introduction

Expansive soil are distributed extensively worldwide (Guney et al. 2007). These soil are highly heterogeneous and unpredictable given their volume change cycle of swelling

H. Shehata and M. Badr (Eds.): *Advancements in Geotechnical Engineering*, SUCI, pp. 144–154, 2021.
https://doi.org/10.1007/978-3-030-62908-3_11

and shrinkage occasioned by environmental and seasonal variations (Dang et al. 2016). For the purpose of foundation design, it is very imperative to carefully recognize and evaluate the expansive soil's capacity to swell or expand (Samtani and Nowatzki 2006) so as to forestall the risk of potential structural failure and resultant economic losses (Sarkar and Islam 2012). The capacity to swell depends on the mineral content of the soil fines or quantity of monovalent cations absorbed on the surface of the clay minerals (Dang et al. 2016). Montmorillonite clays tend to exhibits very high degree of swelling as compared to Illite and Kaolinite which both have moderate to none swell potentials (Samtani and Nowatzki 2006). Expansive soil that exhibit swelling problems consist of silty mudstones, bentonitic mudstones, argillaceous limestone, marls and altered conglomerates (Al-Rawas et al. 2005). Other factors that affects the expansive soil's ability to swell are its relative density, moisture content at compaction, permeability, dry density, location of the groundwater table, past and existing overburden pressure, presence of vegetation and trees, etc. (Khemissa and Mahamedi 2014).

Studies on the use of binders such as cement, lime, fly ash, Ground Granulated Blast furnace Slag (GGBS), Pulverized Fuel Ash (PFA), polymers, polyesters, Cement Kiln Dust (CKD), etc. to improve the consistency of soil of high swelling potential are documented in literatures (Sarkar and Islam 2012), (Al-Rawas et al. 2005) and (Khemissa and Mahamedi 2014). The immediate increase in workability and reduction in plasticity index of the expansive soil after mixing with binders is attributed to flocculation and agglomeration reactions in the hydration process. Meanwhile, the long-term gain in strength is due mainly to pozzolanic reactions (Keramatikerman et al. 2016). Previous studies have used different parameters as design criteria to choose or predict the binder combinations and proportions for stabilization. According to Puppala et al. 2013; soil that possess a Plasticity Index (PI) value greater than 30% may not be used typically for stabilizing or mixing with cement. However, Abbey et al. 2016 have proven in their study that the improvement of soil of natural plasticity index in the range of 37% and 45% by the application of GGBS and PFA to cement can cause a significant reduction in the plasticity index and subsequent strength gain of those soil. Abbey et al. 2016; also showed that the addition of Cement/PFA/GGBS has a significant influence on the PI of soils with lower plasticity than soils with higher plasticity. Khemissa and Mahamedi; demonstrated the effectiveness of the addition of lime and cement in different proportions to predict the swelling pressure and free swelling capacity of soils by using the dry unit weight, natural moisture content, clay content, and the Atterberg limits as basis of assessment. However, the determination of the vertical swell potential and behaviour of expansive soils by the addition PFA and GGBS in different predetermined proportions and combinations to cement treated expansive soil is quite limited in literature. Several methods (direct and indirect) have been developed and applied to measure the swelling potentials of expansive soil (Dang et al. 2016). The indirect methods employed so far involve using soil properties and classification systems to assess the swelling potential, whereas the direct methods deal with the actual physical estimation of the swelling potentials mostly through laboratory schemes Puppala et al. 2016 and Chen 1988. In this study, the application of existing correlations as well as laboratory experiments and numerical analysis on the volume change characteristics of soil improved by the addition of PFA and GGBS to fixed quantities of cement shall be investigated.

2 Experimental Work

2.1 Expansive Soil

In this present study, soil was collected at a depth of about 1.5 m from Al-Andalos area at New Cairo, Egypt and utilized. Natural soil was greenish grey, very hard, laminated silt clay with traces of fine sand and calcareous matters. The ground water table was not being found as shown in Fig. 1.

Fig. 1. Soil structure as site

2.2 Silica Fume

Silica fume, referred to as micro silica or condensed silica fume, is a byproduct material that is used as a pozzolanic as shown Fig. 2.

Fig. 2. Scanning electron microscope micrograph of silica Fume particles.

This byproduct is a result of the reduction of high-purity quartz with coal in an electric arc furnace in the manufacture of silicon or ferrosilicon alloy. Silica fume rises as an oxidized vapor from the 2000 °C (3630 °F) furnaces as listed in Table 1.

Table 1. Chemical analysis of Silica Fume

SiO_2	93.8%	SO_3	0.19%	Na_2O	0.43%
CaO	0.33%	Fe_2O_3	1.48%	TiO_2	0.25%
Al_2O_3	0.36%	MgO	0.41%	Purity	>95%

2.3 Slag

Ground granulated blast-furnace slag as shown in Fig. 3, also called slag cement, is made from iron blast-furnace slag; it is nonmetallic hydraulic cement consisting essentially of silicates and aluminosilicates of calcium developed in a molten condition simultaneously with iron in a blast furnace.

Fig. 3. Scanning electron microscope micrograph of Slag Particles.

The molten slag at a temperature of about 1500 °C (2730 °F) is rapidly chilled by quenching in water to form a glassy sand like granulated material, chemical analysis of slag listed in Table 2.

Table 2. Chemical analysis of Slag

SiO_2	33.28%	SO_3	2.21%	Na_2O	1.2%
CaO	37.1%	Fe_2O_3	3.15%	TiO_2	0.4%
Al_2O_3	13.12%	MgO	7.74	Purity	>70%

3 Laboratory Testing

The soil was placed in an oven at 105 °C for 24 h. The dry soil was pulverized to minus 40 sieve size. Then the soil was mixed with the additives at percentages 1%, 3%, 5%, 6%, 7% and 10% of unit weight of dry soil. Influence of additives on the geotechnical characteristics of expansive soil was investigated by conducting Grain size distribution, Atterberg limits, standard proctor compaction tests, unconfined compressive test, direct shear tests, and swelling tests.

3.1 Physical and Engineering Properties

Soil gradation is determined by analyzing the results of a sieve analysis, or a hydrometer analysis for soils containing appreciable quantities of fine fraction (less than 75 μm).

Since test samples are fine grade soil, hydrometer test was performed to determine grain size distribution of test samples.

In the laboratory, Casagrande's apparatus has been used to determine liquid limit. The recorded values have been plotted on a semi-log scale and approximated as straight line. The water content corresponding to number of 25 has been considered to be the liquid limit (L.L) for the specific soil category under consideration. The plastic limit has been determined for each soil category. The plastic limit (P.L.) has been calculated as the average of three values of water contents for each soil category. For each soil category the plasticity index has been determined as the difference between the liquid limit and plastic limit. Atterberg limit tests were conducted according to ASTM D 4318-87.

3.2 Mechanical Properties

The standard proctor test was used to determine the optimum moisture content of the soil corresponding to a maximum dry unit weight condition. The Test was conducted according to ASTM D 698-78.

Unconfined compression test was carried out on soil specimens. The natural and treated soil were compacted at optimum moisture content and maximum dry density at standard compression test. From molds, specimens of 38.1 mm diameter and 76.2 mm long were extracted and stored in desiccators at room temperature for curing. Samples were tested after 7 and 28 days of curing. Unconfined compressive strength was determined at a loading rate of 1.0 mm/min. The Test was conducted according to ASTM D2166-87.

Direct shear box consist of shear box counter 55 × 55 × 18 mm, applied load which distributed on the total area of specimen through the loading ceil, the direct shear tests were performed with different applied normal stresses. Direct shear test were carried out at a rate of strain of 0.125 mm/min. Vertical and horizontal displacement was recorded by two dial of accuracy 0.002 mm.

According to the free swell method, an initial volume (V_i) of 10 cm^3 of dry soil passing sieve No.40 has been poured into a 100 cm^3 graduate cylinder filled with water. The volume of the swelled soil (V_f) after 24 h has been recorded from the graduation of the cylinder.

The standard oedometer was used to measure swelling potential and swelling pressure. Swelling potential has been used to describe the ability of a soil to swell in terms of volume change, and it is defined as the ratio of increase in height to the original height due to an increase in moisture content. Swelling pressure is designated as the pressure required returning sample to its original volume. Swelling test of samples has been conducted according to ASTM D4546.

4 Results and Discussions

The physical and index properties of the natural soil are summarized in Table 3.

Table 3. Physical and index properties of the natural expansive soil

Soil property	Value	Soil Property	Value
Natural Moisture Content (w_c %)	12	Liquid Limit (L.L %)	90
Specific Gravity (G_s)	2.70	Plastic Limit (P.L %)	32
% of Sand	30	Plasticity Index (P.I %)	58
Passing percentage % (Sieve No.200)	70	Shrinkage Limit (S.L %)	7
% of Silt	10	USCS Classification	CH
% of Clay	60	Activity (A_c) = $\frac{P.I\ \%}{\%\ of\ Clay}$	0.97
O.M.C %	21	Free Swell %* = 60 k $(P.I)^{2.44}$	43
Max. Dry Density (kN/m^3)	15.4	Bulk Density (kN/m^3)	20
Void Ratio (e_o)	0.48	UCS $q_{un.}$ (kN/m^2)	510

* Swell % according to Cartle M. and Bentely, 1991 and k = 3.6×10^{-5}

Figure 4 shows that, all samples can be classified as clay of high plasticity (CH) as per the Unified Soil Classification System (USCS).

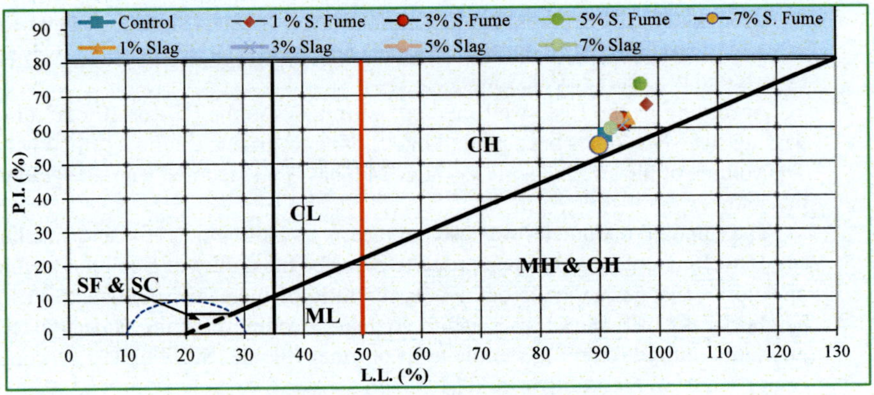

Fig. 4. Effect of additives on soil plasticity.

The addition of Silica Fume and Slag in different proportions to the natural soils resulted in a general unchanged of the soil's liquid Limits and hence the Plasticity Index as seen in Figs. 5 and 6. The most significant change is observed when 7% of Silica Fume was added to soil given that the PI for these mixture proportions was the lowest.

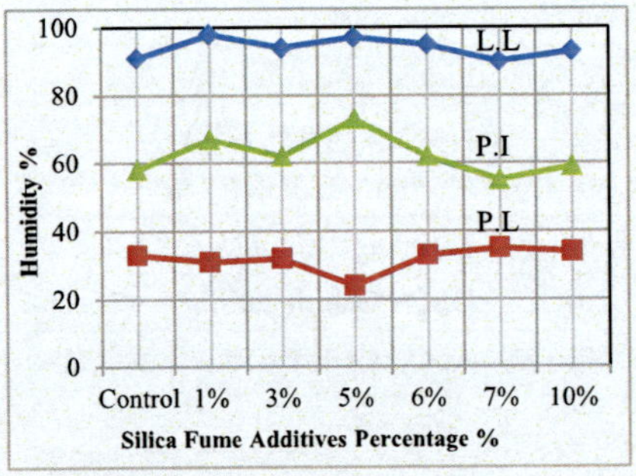

Fig. 5. Effect of Silica Fume additives on liquid limit, plastic limit and plasticity index.

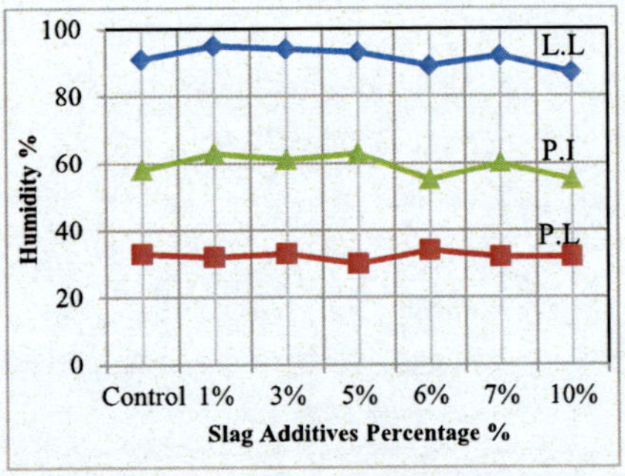

Fig. 6. Effect of Slag additives on liquid limit, plastic limit and plasticity index.

The results of free swell index tests for the untreated and treated soil are shown in Figs. 7 and 8.

According to the free swell method, an initial volume (V_i) of 10 cm^3 of dry soil passing sieve No. 40 has been poured into a 100 cm^3 graduate cylinder filled with water. The volume of the swelled soil (V_f) after 24 h has been recorded from the graduation of the cylinder.

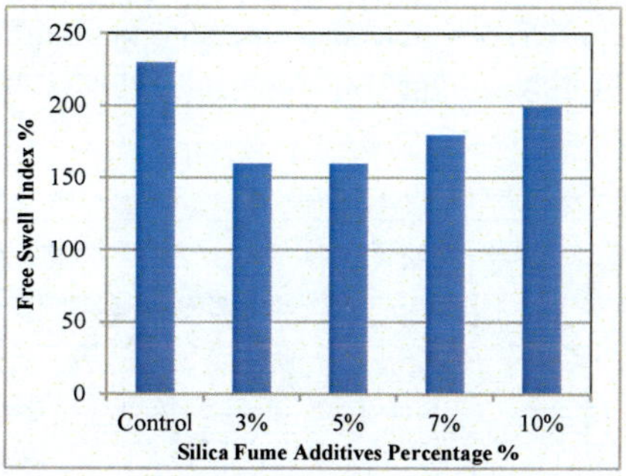

Fig. 7. Effect of S.F additives on free swell index.

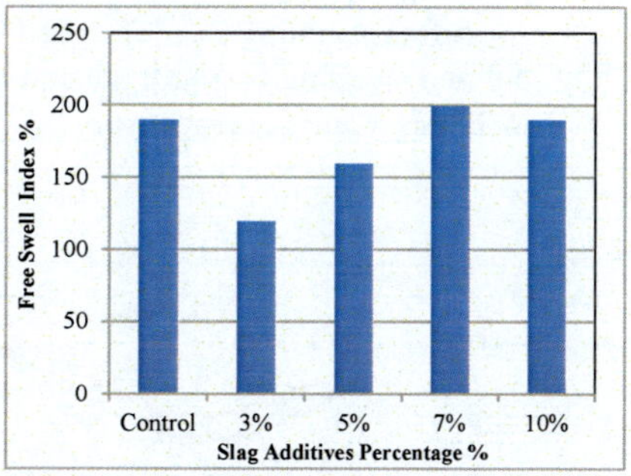

Fig. 8. Effect of Slag additives on free swell index.

Figures 9 and 10, gives the rate of swelling for soil treated by 1%, 3%, 5%, and 7% Silica fume or Slag, in the form of swelling-Time curves and show that, the swelling-Time curves comprise three distinct zones. The initial swelling is generated quickly due to hydration. Next, a high rate of increase in swelling potential is observed in the primary stage followed by a low rate in the secondary stage.

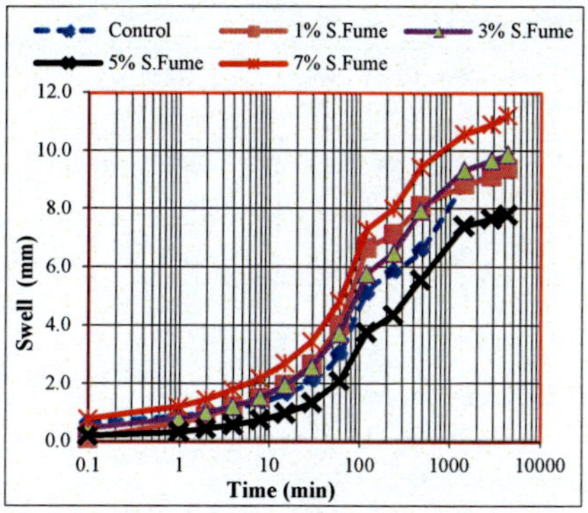

Fig. 9. Swell – Time curves of Silica Fume additives.

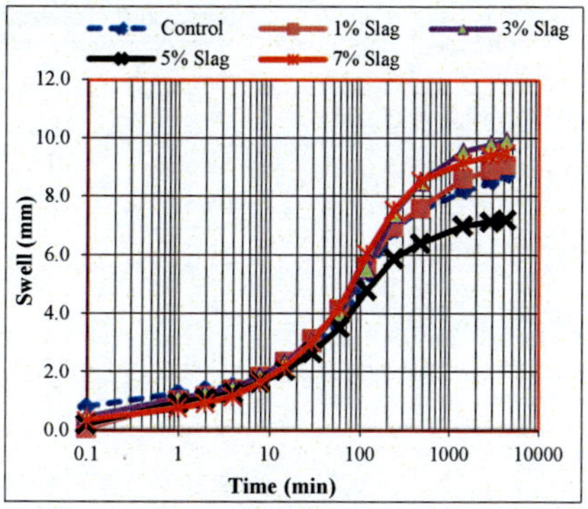

Fig. 10. Swell – Time curves of Slag additives

Results of standard compaction test found that maximum dry density is 15.4 kN/m^3 and its optimum moisture is 21%.

Effect of adding 1%, 3%, 5%, 7% and 10% of Silica Fume and Slag on swelling potential and swelling pressure are shown in Figs. 11 and 12, and result values summarized in Tables 4 and 5, respectively.

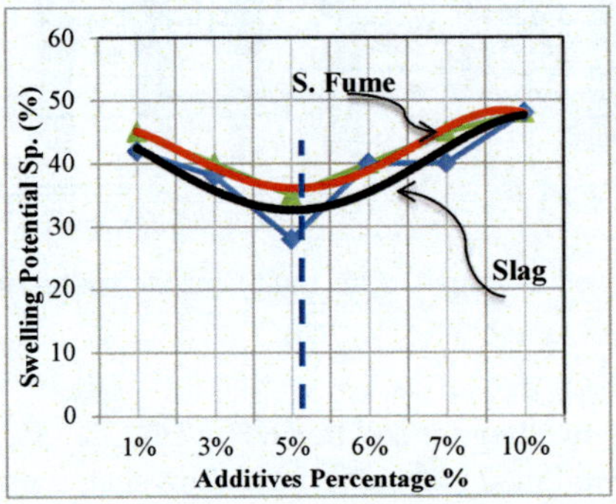

Fig. 11. Effect of Silica Fume and Slag additives on swelling potential.

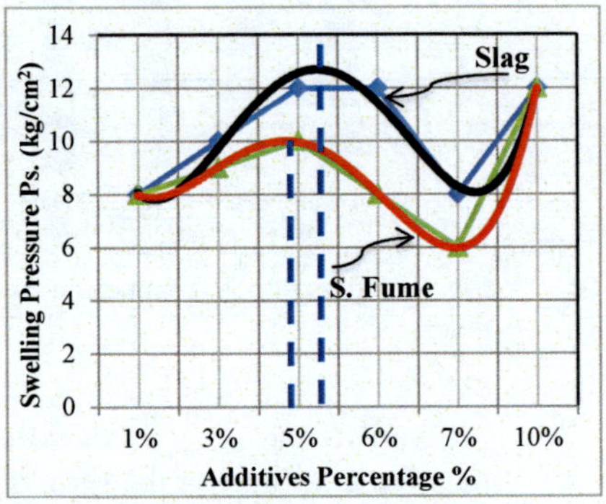

Fig. 12. Effect of Silica Fume and Slag additives on swelling pressure.

Table 4. Effect of additives on swelling potential

Parameter	Swelling Potential (%)						
Additives %	Untreated	1%	3%	5%	6%	7%	10%
Silica Fume	50	45	40	35	40	45	48
Slag		42	38	28	40	40	48

Table 5. Effect of additives on swelling pressure

Parameter	Swelling Pressure (kN/m^2)						
Additives %	Untreated	1%	3%	5%	6%	7%	10%
Silica Fume	600	800	900	1000	800	600	1200
Slag		800	1000	1200	1200	800	1200

Unconfined compressive strength tests were carried out on 7 and 28 days samples of age. Effect of adding 1%, 3%, 5%, 7% and 10% of Silica Fume and Slag on unconfined compressive strengths are given in Table 6.

As can seen in Figs. 13 and 14, Effect of curing time can be sighted from the differences in behavior trend of seven and twenty eight days. 7-days unconfined compressive strength is not greatly impacted by adding silica fume in samples, but its increase by

Table 6. Values of unconfined compressive strength with curing time

Parameter	Unconfined Compressive Strength (kN/m^2)							
Additives %	1%		3%		5%		7%	
Curing time	7-day	28-day	7-day	28-day	7-day	28-day	7-day	28-day
Silica Fume	420	580	360	770	450	720	380	430
Slag	240	420	600	750	300	750	300	600

adding 3% slag. Unconfined compressive strength test results for 28-days show that, the addition of silica fume or slag increase at 3% and straight to 5% then decrease at 7%. The optimal amount of additions for 28-days samples is from 3% to 5% silica fume or slag.

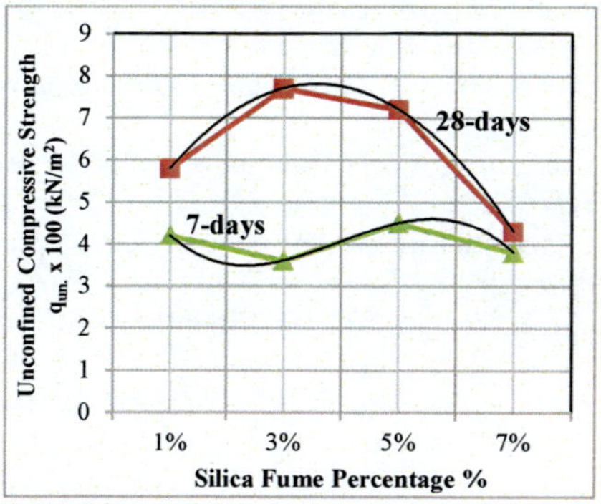

Fig. 13. Effect of Silica Fume additives on unconfined compressive strength.

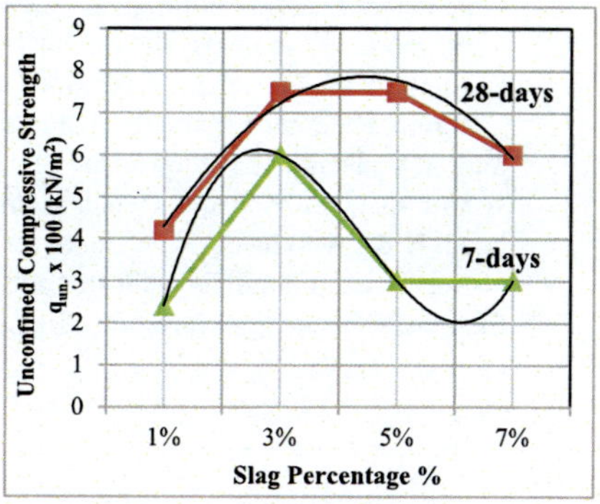

Fig. 14. Effect of Slag additives on unconfined compressive strength.

5 Conclusions

From the study of utilization of both silica fume and slag as improvement aid to reduce swelling potential of an expansive soil, the following conclusions are drown;

1. The expansive nature of soil gets decrease on addition of silica fume and slag.
2. Soil plasticity is not improved with increase of silica fume or slag because of high softness of its particals.
3. The shear strength characteristics of soil are improved on addition of Silica Fume and Slag.
4. Curing time diagrams indicated that because of small size of particals, The optimum amount of additions for 28-days samples is from 3% to 5% silica fume or slag.
5. The utilization of industrial by–products cannot be limited to improvement of strength properties of weak soils only but can be extended to controlling problematic challenges posed by expansive soils.

References

Abbey, S.J., Ngambi, S., Coakley, E.: Effect of cement and by-product material inclusion on plasticity of deep mixing improved soils. Int. J. Civ. Eng. Technol. (IJCIET) **7**(5), 265–274 (2016)

Al-Rawas, A.A., Hago, A.W., Al-Sarmi, H.: Effect of lime, cement and Sarooj (artificial pozzolan) on the swelling potential of an expansive soil from Oman. Build. Environ. **40**(5), 681–687 (2005)

Chen, F.H.: Foundations on Expansive Soils in Mexico. Elsevier, New York (1988)

Dang, L.C., Fatahi, B., Khabbaz, H.: Behaviour of expansive soils stabilized with hydrated lime and bagasse fibers. Procedia Eng. **143**(Ictg), 658–665 (2016)

Guney, Y., Sari, D., Cetin, M., Tuncan, M.: Impact of cyclic wetting-drying on swelling behavior of lime-stabilized soil. Handb. Environ. Chem. **42**(2), 681–688 (2007)

Keramatikerman, M., Chegenizadeh, A., Nikraz, H.: Effect of GGBFS and lime binders on the engineering properties of clay. Appl. Clay Sci. **132–133**, 722–730 (2016)

Khemissa, M., Mahamedi, A.: Cement and lime mixture stabilization of an expansive over consolidated clay. Appl. Clay Sci. **95**, 104–110 (2014)

Puppala, A.J., Manosuthikij, T., Chittoori, B.C.S.: Swell and shrinkage characterizations of unsaturated expansive clays from Texas. Eng. Geol. **164**, 187–194 (2013)

Samtani, N.C., Nowatzki, E.A., Soils and Foundations: Federal Highway Administration Publication, Reference Manual – Volume I (2006)

Sarkar, G., Islam, M.: Study on the geotechnical properties of cement based composite fine-grained soil. Int. J. Adv. Struct. Geotech. Eng. **1**(2) (2012)

Seed, H.B., Woodward, R.J., Lundgren, R.: Prediction of swelling potential for compacted clays. Soil Mech. Found. Div. **88**(3), 53–87 (1962)

SFA. http://www.silicafume.org, Silica Fume Association, Fairfax, Virginia (2000)

Effect of Sand Cushion Thickness and Lateral Extension on Heave of Remolded Swelling Soil

G. E. Abdelrahman[1], Youssef Gomaa Youssef[1], and A. E. Abdeltawab[2](✉)

[1] Soil Mechanics and Foundation Engineering, Civil Engineering Department, Faculty of Engineering, Fayoum University, Faiyum, Egypt
[2] Civil Engineering, Fayoum University, Faiyum, Egypt
aea12@fayoum.edu.eg

Abstract. Swelling soil is one of the most common problematic soils. Swelling phenomenon can cause severe damage to the overlying lightly loaded structures. Many alternatives have been proposed to mitigate the effects of expansive soil on civil foundations such as pre-wetting, sub-drainage, hydraulic barriers, chemical stabilization, and mechanical treatments. Sand replacement is considered to be one of the most effective and practical techniques that can be utilized for the construction of safe foundations. Experimental study using a large-scale box model is used to investigate the effect of sand cushion on reduce the heave of the expansive soil. In addition, Parametric study was performed to assess the effect of sand cushion thickness and lateral extension on the heave of the footing. The results demonstrated that the use of sand cushion reduces the heave of the footing by increasing the thickness of the sand cushion. Moreover, the rate decrease for sand cushion thickness greater than two-thirds of the footing width. Thus, the optimum thickness of the sand cushion is 2/3 footing width. The footing heave decreases with increasing the lateral extension, *Le*, of sand cushion, and the rate of reduction of heave becomes negligible when using lateral extension greater than twice thickness of sand cushion.

1 Introduction

Many buildings, pavements, airport runways, retaining walls, dams, and bridges are distressed by swelling soils. These natural high plasticity soils typically contain clay mineral montmorillonite that exhibits high swelling with the increase of water content (Chen 1988). These types of soils are mostly found in arid and semi-arid climatic regions. Many locations in Egypt are well-known by the presence of swelling soils such as; Nasr – City, Cairo – Suez Road, El – Fayoum City, Kom Ombo City, Aswan City, New Valley, El – Sadat City, and Sohag – Safaga highway (Elbeih and Soliman 2015). With these soils experiencing volume changes, shallow foundation may exhibit surficial distress due to the swell-shrink behavior which leads to cracks in these structures.

Sand replacement is considered to be one of the most effective and practical techniques that can be utilized for the construction of safe foundations (Awad 2005). This method is used because of its many advantages such as the sand replacement can be compacted to a higher density, which leads to higher strength, does not require specialized

H. Shehata and M. Badr (Eds.): *Advancements in Geotechnical Engineering*, SUCI, pp. 155–163, 2021.
https://doi.org/10.1007/978-3-030-62908-3_12

equipment, and it requires no soil additives. However, it does have some disadvantages such as the high volume of excavation may be costly, even after excavation, the underlying and adjacent expansive soils often must be protected by horizontal or vertical membranes.

There is a lot of researches that has examined and investigated the use of sand to control the heave of the swelling soil such as (Ahmed 1985) concluded that providing isolated footing with a sand cushion as a good practice to minimize the swelling pressure as it is possible to reduce the swelling pressure by 25% when using sand replacements with a thickness equal to one-fifth of the footing width.

El-Nahhas et al. 1998a, b studied the effects of the magnitude of active depth, *Ha*, and the thickness of soil replacement on the heave of swelling soil. The results of tests show if 40% of the maximum active thickness of an expansive layer is removed and replaced with compacted non-expansive soil, the maximum heave would be only 54% of that anticipated with no soil replacement.

Awad 1993 concluded that the reduction of final footing heave in case of treated expansive soil using sand replacement is 30% and 52% when the sand thickness is half and one of the footing widths, respectively.

Youssef et al. 2010 investigated the effect of sand cushion parameters on the swelling behavior of expansive soils subjected to changes in soil suction profiles due to climate variations by using a numerical model. The analysis presented provides considerable insight into the effect of the sand cushion as a mitigation method on the behavior expansive soils. Conclusions may be summarized as follow: sand cushion depth has a significant effect on decreasing footing heave and the lateral extension of the sand cushion has a significant effect on the heave of footing. However, increasing lateral extension more than twice the depth of sand cushion is insignificant on footing heave.

Treatment of expansive soil beneath isolated footing using sand replacement is considered one of the most effective, and practical techniques that can be utilized for the construction of safe shallow foundations. The objective of this research is to investigate the effect of sand cushion on heave of the footing models resting on untreated and treated expansive soil Also, finding suitable and economic thickness of replacement sand over a layer of swelling soil was one of the most important objectives of this research.

2 Materials

2.1 Sand Cushion

Fine to medium sand soil is used as layers of sand cushion. Laboratory tests such as sieve analysis test, specific gravity test, compaction test, and direct shear test were conducted to assess sand properties. Table 1 show the sand properties.

Table 1. Sand cushion properties

Properties	Values
Bulk unit weight (KN/m^3)	14.8
Specific gravity, *Gs*	2.65
Fine (%)	1.16
Coefficient of Uniformity, C_U	3.33
Coefficient of Curvature, C_C	1.57
Unified Soil Classification	SP
Optimum Moisture Content, *O.M.C*, (%)	11
Max-Dry Density, *MDD*, (KN/m^3)	1.86
Internal Friction Angle, φ	34

2.2 Expansive Soil

In this study, the swelling soil is remolded from 40% bentonite and 60% kaolinite clay. The laboratory tests carried out on expansive soil were Atterberg limits, hydrometer test, compaction test, free swell test, specific gravity, and swelling pressure test. The soil used is classified as high plasticity clay (CH) according to Unified Soil Classification System. The swelling soil properties are shown in Table 2.

Table 2. Swelling soil properties

Properties	Values
Specific Gravity, *Gs*	2.57
Silt Content (%)	36
Clay Content (%)	64
Liquid Limit, *L.L*, (%)	134
Plastic Limit, *P.L*, (%)	34
Plasticity Index, *P.I*, (%)	100
Shrinkage Limit, *Sh.L* (%)	12
Optimum Moisture Content, *O.M.C*, (%)	21
Max-Dry Density, *MDD*, (KN/m^3)	16
Free Swell, *F.S*, (%)	350
Swelling Pressure, *Ps*, (kpa)	260

3 Experimental Work

A large-scale box model is built to measure the heave of footing due to expansive soil. The dimensions of experimental model are shown in Fig. 1. Also, photo of the model at the laboratory is shown in Fig. 2. A square steel block with dimensions of 100 × 100 × 10 mm was used to simulate a rigid footing condition.

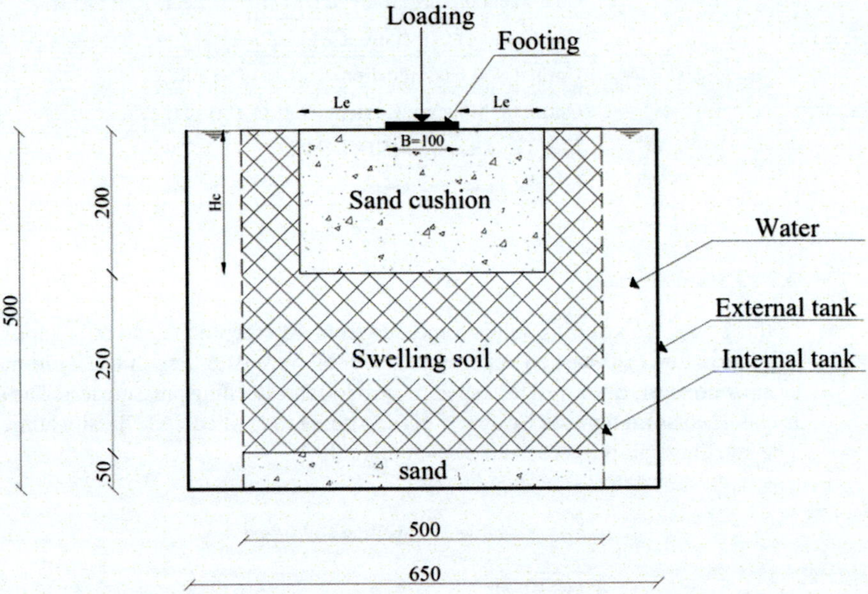

Fig. 1. Experimental model sketch

Several experiments were conducted to investigate the effect of the thickness and width of sand cushion on the footing heave. The soil is placed in the model and compacted on layers to the maximum density. Then the footing is placed and loading begins with the filling of the external tank with water. The heave of footing is measured using two dial gauges until insignificant movement is observed. The soil specimen begins to absorb water from the holes in the perforated plate. The absorbed water enters between the clay platelets causes soil to swell. The footing heave is measured for this method and axial swell is calculated.

Sand bed thickness was maintained constant as 50 mm in all tests, and the swelling soil thickness, *Hs*, was constant as 250 mm. Sand cushion and expansive clay have been compacted to max dry density. Weight loads of 50 kg (50 kPa) were applied on the footing. Sand cushion thickness, *Hc*, of 0, 1/3*B*, 2/3*B*, and *B* are used when studying the effect of sand cushion thickness on the heave of footing where *B* is the footing width. Also, lateral extension, *Le*, of the sand cushion is used equal to *Hc*, 1.5*Hc*, 2*Hc*, 2.5*Hc*, 3*Hc*, and 4*Hc* to assess impact of lateral extension of sand cushion on the heave of footing, where Hc is the sand cushion thickness.

Fig. 2. The large-scale box model at the laboratory

4 Test Results and Analysis

The aim of these tests is to study the effect of thickness and lateral extension of sand cushion layers resting on expansive soil on the footing heave. When the test begins and water enters the internal model from its sides, the space between the external tank and the perforated plate of the internal tank is filled with water. In the next sections, each test was performed, finding the footing heave, and then analyzing these results carefully.

4.1 Effect of Sand Cushion Thickness

The effect of variation of sand cushion thickness, *Hc*, on the heave of footing was studied. It is varied as follows: 0, 1/3*B*, 2/3*B*, and *B*, where *B* is the width of the footing. The relationship between time and axial swell, *A.S*, for swelling soil using a sand cushion is shown in Fig. 3. Axial swell, *A.S*, is equal to the ratio of footing heave to swelling soil thickness ($\Delta H/Hs$). Axial swell is 6.04%, 4.308%, 3.468%, and 3.10% fr the sand cushion thickness are equal to 0, 1/3*B*, 2/3*B*, and *B*, respectively.

It can be found that the footing heave decreases with the thickness of the sand cushion increasing. It's obvious that within the first six days of testing the rate of swelling is much higher than it started to decrease. By the end of the eleventh day, the rate began reaching a negligible value. The maximum axial swell, *A.S*, is 6.04% for swelling soil without sand cushion, while it reduced when using a sand cushion to reach the minimum value

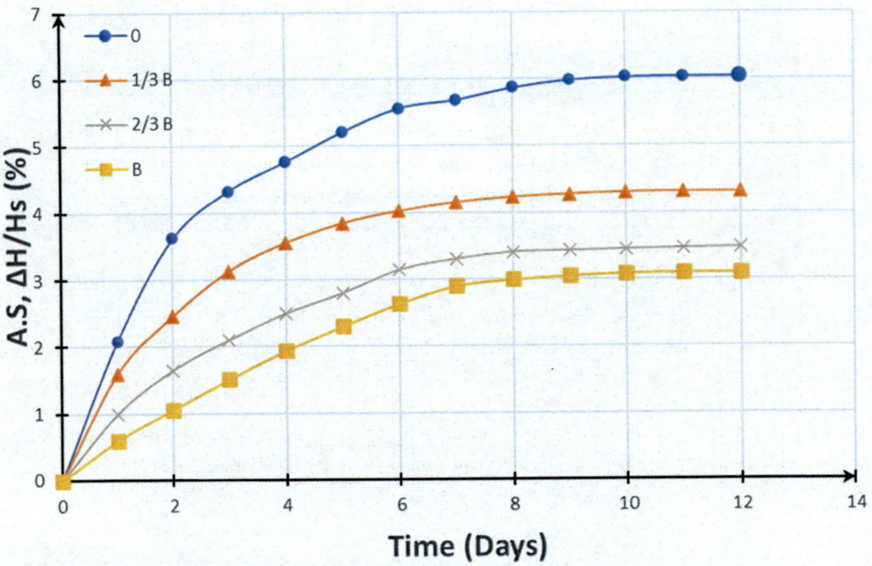

Fig. 3. Time-axial swell curves for variation thickness of the sand cushion

of swelling potential is 3.10% when used sand cushion thickness is equal to the width of footing. Thus, the maximum reduction percentage was 48.68% of the footing heave. This reduction percentage reduces with decreasing sand cushion thickness to the minimum value of 28.67% at a thickness of 1/3*B*. Also, it was found that the heave reduction rate decreases with increasing thickness of the sand cushion.

There are many previous studies that deal with the use of sand cushion to reduce footing heave.

Table 3 shows a comparison of the percentage of reduction in heave between this research and previous researches. From this table, it can be said that with the different types of swelling soil used and the way tests are conducted in these studies, however, the reduction percentage in the heave is very similar in all these studies. For example, in the case of using sand as replacement soil over swelling soil with a thickness equal to the width of the footing, the percentage of reduction in heave was equal to 48.68, 52, and 54 for this study, (Awad 1993), and (El-Nahhas et al. 1998a, b), respectively. Figure 3 illustrates the obtained results for different sand cushion thickness versus the obtained heave of footing. These data showed the same trend obtained by, (El-Dean et al. 2016).

4.2 Effect of the Lateral Extension of Sand Cushion

The effect of sand cushion width on the footing heave has been studied by changing the lateral extension, *Le*, of sand cushion to be *Hc*, 1.5*Hc*, 2*Hc*, 2.5*Hc*, 3*Hc*, and 4*Hc*. The lateral extension, *Le*, is defined as the extending distance of sand cushion from the boundaries of the footing in each side as shown in Fig. 1. The sand cushion thickness, *Hc*, remains constant and is equal to 0.5B. Figure 4 shows the time-axial swell curves

Table 3. Comparison with previous research

References	Sand cushion thickness, *Hc*	Reduction of footing heave (%)
This study	1/3 B	28.67
	2/3 B	42.58
	B = 0.40 Ha	48.68
(El-Nahhas et al. 1998a, b)	0.40 Ha	54
(Awad 1993)	0.50 B	30
	B	52
(Abouleid and Reyad 1985)	0.38 B	31%
(Ahmed 1985)	0.20 B	25%

Note: B = width of footing, Ha = active depth of swelling soil.

for the variation of lateral extension, *Le*, of a sand cushion. In addition, the values of the maximum *A.S* for each case are illustrated in Fig. 5.

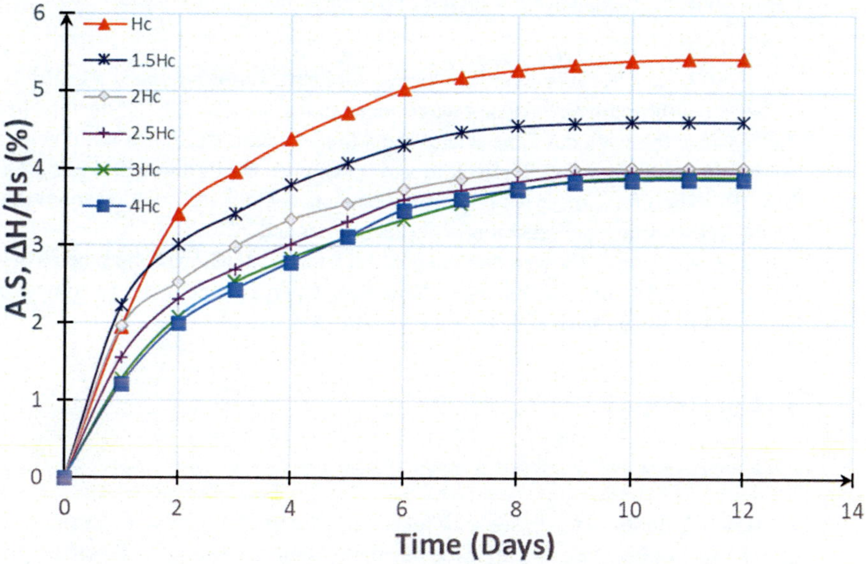

Fig. 4. Time-axial swell curves for variation of lateral extension of sand cushion

Based on Fig. 4 and Fig. 5, the lateral extension, *Le*, of the sand cushion has a significant effect on the heave of footing. Also, it can be observed that the heave of the footing decreases with increasing the lateral extension, *Le*, of a sand cushion. The maximum axial swell value is equal to 5.426% for lateral extension is equal the sand cushion thickness, while this value reduced to reach the minimum value is 3.868% with

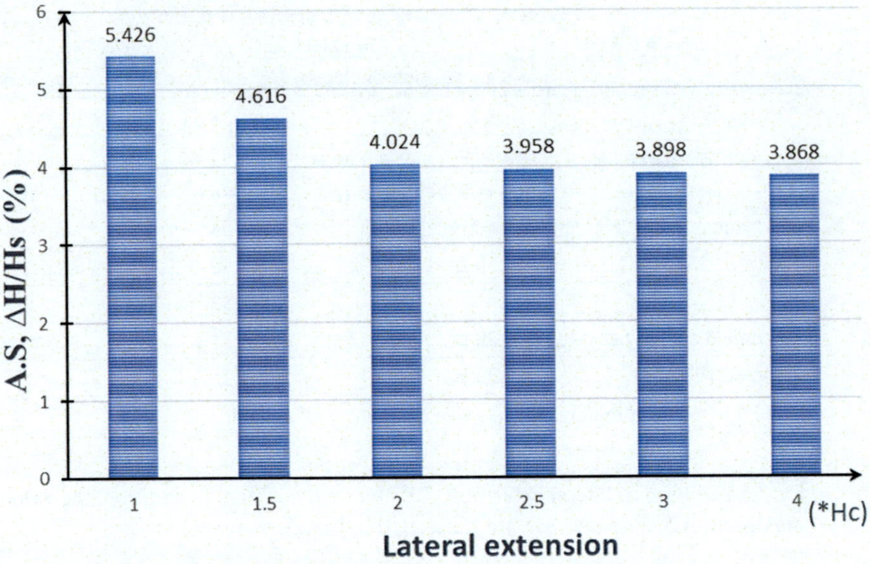

Fig. 5. Maximum axial swell for variation of sand cushion lateral extension

increasing *Le* to 4*Hc*. It is clear that the rate of heave started to reach a negligible value by the end of the tenth day. Also, it was noticed that more than 90% of the total heave was reached after the sixth day. The rate of reduction of heave decreases with increasing the lateral extension, *Le*, of sand cushion, and it becomes negligible when using *Le* greater than 2*Hc*. Therefore, increasing lateral extension, *Le*, of sand cushion more than twice the thickness of the sand cushion is not recommended.

The current study showed that the optimal lateral extension of the sand cushion layer is twice the sand cushion thickness. This experimental conclusion is compatible with the results of the numerical study by (Youssef et al. 2010).

5 Conclusions

From analysis of results, the following conclusion may be drawn as:

1. Heave of footing decreases with increasing the thickness of sand cushion, *Hc*.
2. The rate of decrease in the heave decreases with increasing sand cushion thickness.
3. The maximum axial swell ($\Delta H/Hs$) is 6.04% for swelling soil without sand cushion, while it reduced when using a sand cushion to reach the minimum axial swell is 3.10%.
4. The maximum reduction percentage of footing heave was 48.68% when using the sand cushion thickness equal to the width of footing.
5. The lateral extension, *Le*, of sand cushion has a significant effect on the heave of footing.

6. The rate of reduction of heave decreases with increasing the lateral extension, *Le*, of sand cushion, and it becomes negligible when using lateral extension greater than 2*Hc*.

References

Abouleid, A.F., Reyad, M.M.: Study of the behaviour of isolated footings on expansive soils. In: Proceeding of The Eleventh International Conference on Soil Mechanics and Foundation Engineering, San Francisco, 12–16 August 1985. Publication of: Balkema (AA) (1985)

Ahmed, A.A.: Study on the behavior of some Egyptian expansive soils. thesis, Faculty of Engineering, Ain Shams University (1985)

Awad, T.: Effect of expansive soil on imposed stress of footing and supper structure. Thesis, Faculty of Engineering. Ain Shams University (1993)

Awad, T.: Properties of expansive soil treated with sand-lime cushion. Tenth International Colloquium on Structural and Geotechnical Engineering, Faculty of Engineering, Ain Shams University, vol. 16, 1–12 (2005)

Chen, F.H.: Foundations on Expansive Soils. In: Found on Expansive Soils (1980)

El-Dean, M.R.G., Fouad, M.T., Bahr, M.A.: Characteristics of Some Unsaturated Soils Based on Field and Laboratory Tests, thesis, Eng. Mohamed Ragab Gamal El-Dean Under the Supervision of Dr. Mohey Elmashaad, Faculty of Engineering Civil Engineering Department (2016)

El-Nahhas, F., Awad, T., Abd-Moatty, N.: Prediction of Ground Heave due to the Swelling of Multi-Layered Soil Deposits. In: 8th International Colloquium on Structural and Geotechnology Engineering, Faculty of Engineering, Ain Shams University (1998)

Elbeih, S.F., Soliman, N.M.A.: An approach to locate and map swelling soils around Sohag-Safaga road, Eastern Desert, Egypt using remote sensing techniques for urban development. Egyptian J. Remote Sens. Space Sci. **18**(1), S31–S41 (2015). https://doi.org/10.1016/j.ejrs.2015.07.002

El-Nahhas, F., Awad, T., Abdel-Moatty, N.: Prediction of Ground Heave Due to Swelling of Multi-Layered Soil Deposits. Eighth International Colloquium on Structural and Geotechnical Engineering 15–17 December, L99g Ain Shams University Cairo, Egypt (1998)

Youssef, Y.G., Darrag, A.A., Elkady, T.Y.: Numerical Modeling of Expansive Soils Using Numerical Modeling of Expansive Soils Using. Ph.D. (2010)

Dynamic Response of Stabilized Soft Clay Under Cyclic Loading at Low Shear Strain Level

Ahmed M. Karim[1](✉), Osama M. Ibrahim[2], and Mohamed I. Amer[3]

[1] Faculty of Engineering, Misr University for Science and Technology, 6th of October, Egypt
ahmed.karim@must.edu.eg
[2] High Institute for Engineering at 15th May City, Cairo, Egypt
[3] Faculty of Engineering, Cairo University, Giza, Egypt

Abstract. Grand construction projects all over Egypt require large infrastructure projects. In Northern Egypt where large buildings are needed for new developments, large areas of soft soils are abundant. These buildings transfer heavy static and dynamic loads to soil. Soft clays can be stabilized to carry these loads with minimum settlements and high shear strength. This research focuses on improving the static and dynamic properties of soft clay using chemical stabilization by cement or lime. undisturbed samples were collected from Belkas, Dakahleia – Egypt. The engineering properties were determined and classified as soft silty clay (CH). Low proportions of added cement or lime between 4 and 8% by weight were mixed with the natural samples, The effect on the strength and dynamic properties of reconstituted samples were investigated. Three series of a free-fixed type of resonant column tests were conducted in torsional mode on natural soft silty clay, stabilized soft silty clay with cement, and stabilized soft silty clay stabilized with lime. The influence of confining pressure and stabilizer amount on dynamic shear modulus (G_{max}) and material damping (D_{min}) for natural and treated samples were studied. The threshold shear strain γ_{th} for each treatment type is investigated at small-to-medium shear strain amplitude levels to investigate the effect of torsional shearing on material softening behavior and normalized shear modulus (G/G_{max}) degradation of treated soil. The results are presented and discussed displaying percentages stabilizers that led to significant improvement in the soil stiffness and shear modulus using cement or lime as the stabilizers, The damping ratio, (D_{min}) decreased by increasing stabilizers dosage.

Keywords: Soft silty clay · Chemical stabilization · Lime · Portland cement · Resonant column · Maximum shear modulus · Threshold strain · Minimum damping ratio

1 Introduction

Soft clay soils in their natural state may not be suitable for supporting structures at construction sites always. The characterization of such soils and the determination of their properties with reliability for use in foundations can be even difficult (Mezni and

H. Shehata and M. Badr (Eds.): *Advancements in Geotechnical Engineering*, SUCI, pp. 164–174, 2021.
https://doi.org/10.1007/978-3-030-62908-3_13

Bouassida 2018). In such cases, a soil may be improved to increase its bearing capacity, decrease the settlement, and meet the construction requirements. In Egypt, large areas of soft clay deposits are widely spread in Northern of the Nile Delta. Chemical stabilization of soft soils has been extensively used in both shallow and deep applications to improve the engineering soil properties such as strength and deformation behavior (Bergado et al. 1996; Chen and Wang 2006; and Ho and Chan 2011). Treatment of upper 3.0 to 5.0 m of soft clay by mixing the soft clay with a stabilizer agent was reported to give improvement in physical properties of the soil, and to increase its bearing capacity (FHWA 1998).

Cement stabilization is one of the practical stabilization techniques for silty clayey soils expect those containing organic matters (Ramaswamy et al. 1984). When cement is mixed with the silty clay soils, a chemical reaction occurs between cement and soil particles. This phenomenon is due to formation of calcium silicate hydrate and calcium aluminate hydrate (Moh 1962). As cement hydrate, a gel is formed first then hardens. Upon hardening, a cellular material that encapsulates the soil particles is formed associated with forming strong bond between the soil particles, thus producing a hard and durable material (Winterkorn and Pamukcu 1991). Although cement hydration occurs immediately when water is added, secondary reactions are slower and may go on for several months. The typical proportions of cement to be added are (5–15%) of the weight of the soil (Jaritngam and Swasdi 2006). Generally, the optimum proportion is determined experimentally by performing compressive strength tests. Efficiency of mixing in-situ soil may lead to lower cement concentration in soils (Punmia 1994). Table 1 displays typical percentages of cement by weight that can be tested initially for different soil types (Rodriguez et al. 1988).

Table 1. Percentages of cement to be tested initially for soils (Rodriguez et al. 1988)

Type of soil	Percentage of cement by weight
GW, GP, GM, SW	3–8
SC, GC	5–9
SP, SM	7–11
ML	7–12
CL, OL, MH	8–13
CH	9–15
OH	10–16

The compressive strength of cement treated clay increased with the increase of curing time (Kawasaki et al. 1981). The final compressive strength of the stabilized clay increased with the increase of cement content (Uddin et al. 1997).

Lime reacts with clay minerals of the soil; this reaction forms insoluble gel of calcium silicate, which cements the soil particles. The cementing agent is similar in nature to that of cement. The difference is that in the case of cement, the calcium silicate gel is formed by hydration of calcium silicate in the cement itself, but with lime, the gel is formed only after removal of the silica from clay minerals due to reaction with lime. The silica gel coats and binds clay lumps in the soil pores (Ingles 1970).

Short-term reactions include hydration for quicklime and flocculation (ion exchange). Long-term reactions are cementation and carbonation. Quicklime reacts with the water in the soil. During the placement of lime columns and layers, the heat generation and expansion of the lime further enhance the consolidation effect. When lime is mixed with clay, sodium and other cations adsorbed to the clay mineral surfaces are exchanged with calcium, this action makes the clay minerals connected together. Lime causes reduction in clay plasticity index, making it more workable and increasing its strength (Winterkorn and Pamukcu 1991).

The deformation parameters on the strain range of 10^{-6} to 10^{-2} are important design parameters. In particular, the strains operating in the ground that affects the serviceability limit state of structures are usually smaller than about 5×10^{-3} (Biarez et al. 1999).

It has been revealed that exposing the soil to very small strains (generally until around 10^{-5}), leads to an elastic behavior, and then, followed by a hysteretic and non-linear behavior with increasing the strain level, where the stiffness decreases. In practice, it is therefore important to evaluate the stiffness of soil for the stress and strain states that are relevant for a specific a geotechnical design (Cavallaro et al. 2001).

2 Experimental Apparatus and Procedure

In this research study, a testing program was performed on reconstituted samples of soft silty clay stabilized with either cement or lime. The tests were performed to get the physical and dynamic properties of the soil at its natural state and after adding different percentages of cement or lime on reconstituted samples.

2.1 Natural Soil

Undisturbed Shelby tube natural soft silty clay specimens were extracted from Belkas-Dakahlyia at 5.0 m depth by the Foundation and Soil Mechanics Research Laboratory, Faculty of Engineering Cairo University. The soil was prepared to perform classification and strength tests to determine the physical and mechanical properties of the natural soft soil according to the Egyptian standards.

2.2 Mixtures

To produce the mixtures implemented in this study, the natural samples were oven dried for 24 h, then finely divided and mixed with cement or lime mixture. Three mixtures were prepared per additive percentage of 4, 6, and 8% by weight of dry soil. The reconstituted samples have diameter of 50 mm and length of 100 mm, and were cured for a period of 7 days in a humid container.

2.3 Resonant Column Apparatus and Testing Sequence

The resonant column test (RCT) is one of the most accurate and repetitive ways of determining the small strain shear modulus (Lo Presti et al. 1993).

Resonant column apparatus was used to investigate dynamic properties for each group. The resonant column test is performed by applying a sinusoidal oscillation via an electromagnetic drive system. The four pairs of coils are connected in series so that a net torque is applied to the sample during torsional tests as indicated in Fig. 1.

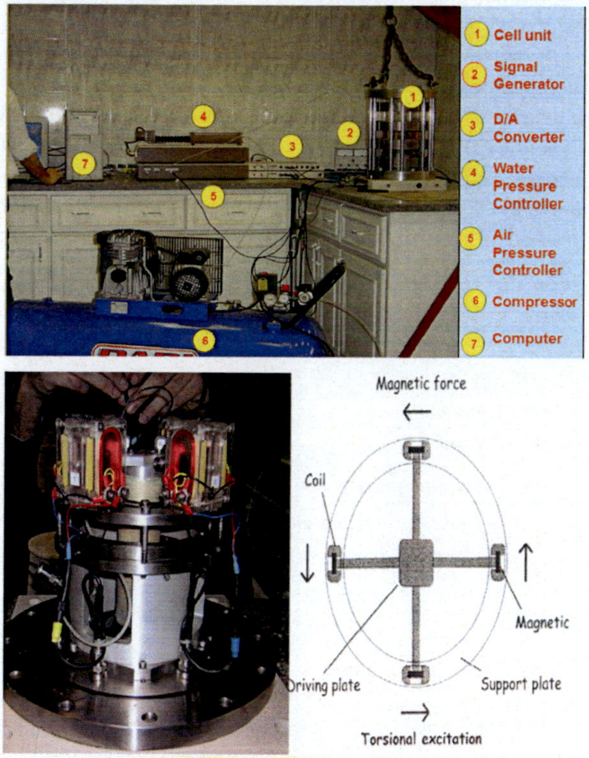

Fig. 1. Resonant column apparatus system

After 7 days of curing in a humid container, the specimen is assembled into the RC apparatus and the different RC components were set into place. An initial isotropic confining pressure, σ'_0, of 25 kPa was applied. A series of RC tests were performed at time intervals of 0, 1, 2, 3, 4, and 6 h elapsed from the time that the initial pressure was applied and was kept constant along test period. All RC tests were performed by sending a low-amplitude (100 mV peak to peak) sinusoidal signal from a dynamic signal generator power supply to the torsional driver fixed on top of the specimen. The frequency of the signal is incrementally changed by sweeping the frequency scale until the frequency response curve is obtained, which allows determination of linear shear modulus G_{max}

and damping ratio D_{min}. Following the last RC test completion, σ'_o is increased from 25 kPa to 50 kPa, and another series of RC tests were performed at the same elapsed times. The same test procedures is repeated for $\sigma'_o = 100$ kPa.

3 Experimental Program

3.1 Characterization of Natural Soil

Tests were conducted on saturated soft silty clay obtained from Belkas, Dakahleia governorate, Delta region of Egypt to determine its physical and mechanical properties before and after mixing with either cement or lime. Visual classification and field preliminary examinations from boreholes indicated the existence of the soft soil in this site. The grain size distribution of the natural soft silty clay indicated that the soil is composed of 18.0% silt, and 82.0% clay. The natural soft clay has a liquid limit of 82%, plastic limit of 20%, and accordingly its plasticity index is 64%. The field natural water content was determined to be 58%. According to the Unified Soil Classification System (USCS), the soil is classified as high plasticity clay (CH). The above and other physical properties of the soft clay are summarized in Table 2.

Table 2. Physical properties of natural soil

Property	Value
Bulk unit weight (gm/cm^3)	1.62
Specific gravity	2.72
Natural water content (%)	58
Liquid limit (LL) %	82
Plastic limit (PL) %	20
Plasticity index (PI) %	62
Degree of saturation (S) %	95
Silt content %	18
Clay content %	82
consistency index	0.39
Unconfined compressive strength (kPa) (kg/cm^2)	37
USCS: Unified Soil Classification	

3.2 Effect of Stabilizer Percentage on Unconfined Strength

The unconfined compressive strength tests were performed on reconstituted samples treated with cement or lime at percentages of 4, 6, and 8% by weight. The unconfined compressive test is performed after 7 days curing period. Table 3 displays the obtained strength values at each stabilizer percentage.

Table 3. Unconfined compressive strength for treated soil

Additive percent	Unconfined compressive strength (kPa)	
	Cement stabilization	Lime stabilization
4%	120	50
6%	>30	80
8%	>30	140

3.3 Dynamic Response Tests

Two main goals were aimed at in this study. The first is related to very small strain properties, tested with the resonant column apparatus, where major concerns are centered on the determination of maximum shear modulus (G_{max}), and minimum damping ratio (D_{min}). The second goal is concerned with the effects of degradation on the shear modulus and damping ratio measured with various types of cyclic loading test apparatus and covers a larger strain level.

3.3.1 Linear Dynamic Response

Linear models imply that the strain will always return to zero after cyclic loading, and since a linear material has no limiting strength, failure cannot occur. The linear dynamic response is investigated at very low strain (less than 10^{-5}) by sending a low-amplitude (100 mV peak to peak) sinusoidal signal in the RC apparatus. The normal and treated samples are tested under confining pressures (σ'_o) of 25, 50, 100 kPa respectively. The tests were performed after exposing the sample to these confining pressures for 6 h. In the stage of linear dynamic behavior, the focus was on determining the Max. shear modulus (G_{max}) and the minimum damping ratio (D_{min}) at this confining pressure and strain level. By increasing additive (cement or lime) percentage by weight and or the confining pressure (σ'_o), the material stiffness increased, which led to increasing in the value of shear modulus (G_{max}) and decreasing in the value of damping ratio (D_{min}) as shown in Figs. 2, 3, and 4.

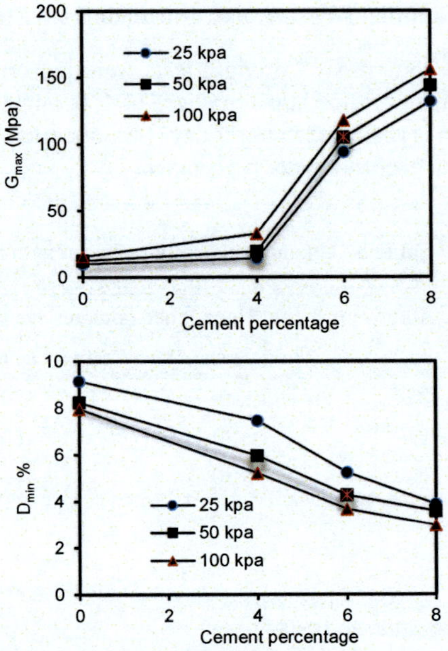

Fig. 2. Effect of cement percentage by weight & confining pressure on shear modulus (G_{max}) and damping ratio (D_{min}).

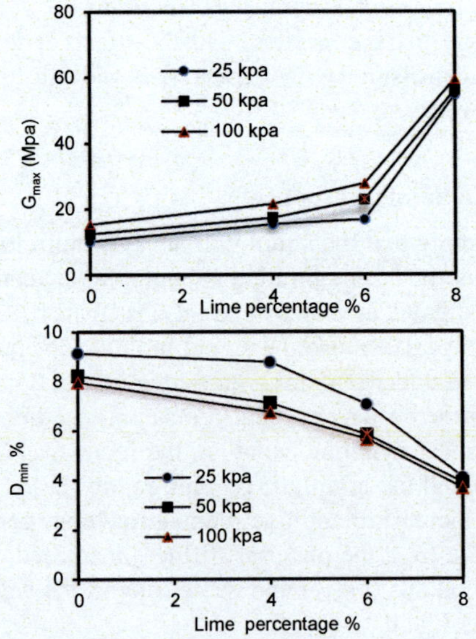

Fig. 3. Effect of lime percentage by weight & confining pressure on shear modulus (G_{max}) and damping ratio (D_{min}).

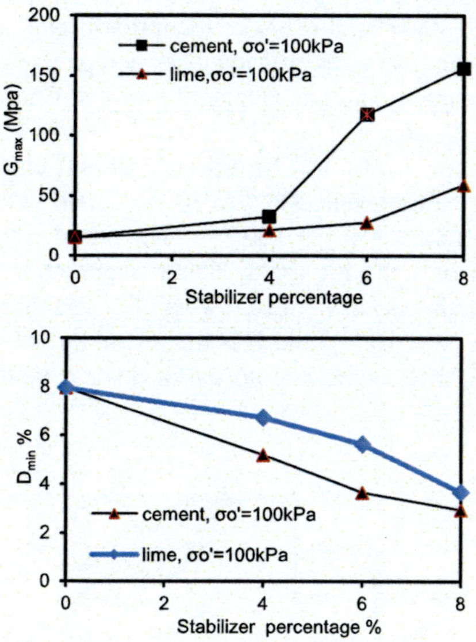

Fig. 4. Effect of stabilizer percentage by weight on shear modulus (G_{max}) and damping ratio (D_{min}) at $\sigma'_o = 100$ kPa.

3.4 Non-linear Dynamic Response

The non-linear stress-strain behavior of soils can be represented more accurately by cyclic non-linear models that follow the actual stress-strain path during cyclic loading. Shear modulus of soils is considered to be maximum at the shear strain level below 10^{-5}, whereas damping ratio of soil is minimum within this strain range. As shearing strain increases, soil shear modulus decreases but damping ratio increases, and a stiffness degradation occurs and the cyclic loading behavior changes into a hysteretic and non-linear response to loading. The shear strain that the soil material begins to show significant stiffness degradation is defined as **threshold shear strain** (γ_{th}). In this work the threshold shear strain is determined for several specimens as shown in Table 4.

Table 4. Threshold shear strain values for treated soil

Additive percentage	Cement stabilization	Lime stabilization
4%	1×10^{-5}	3×10^{-5}
6%	4×10^{-6}	2×10^{-5}
8%	2×10^{-6}	1×10^{-5}

The threshold shear strain for the natural soft clay sample is determined equal to 7×10^{-5}. Accordingly, increasing stabilizer percentage caused a reduction in threshold shear strain value due to the improvement in soil stiffness and reduction in ductility.

This study goals also included the determination of characteristics of the shear modulus (G) and damping ratio (D) for the natural soft clay and treated samples at small to medium shear strain levels (10^{-5} to 10^{-2}). Small to medium shear strain amplitude RC tests were conducted at σ'_o=100 kPa. The purpose is to investigate the shear modulus degradation of soil with strain. Hence degradation of normalized shear modulus (G/G_{max}) with strain was investigated where G_{max} is the initial shear strain modulus for very small strain ($\gamma < 10^{-5}$), and G is the secant modulus of soil.

Upon the RC test results, the normalized shear modulus (G/G_{max}) is displayed in Fig. 5.

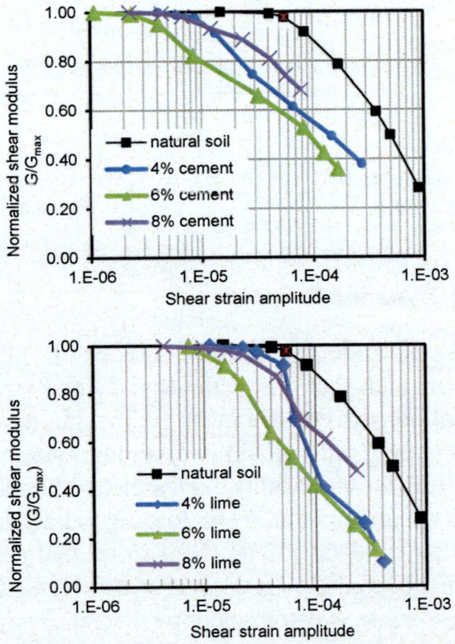

Fig. 5. Effect of cement and lime percentage by weight on shear strain amplitude modulus (G_{max}) and damping ratio (D_{min}).

4 Conclusions

RC laboratory tests were performed on field samples before and after stabilization with either cement or lime as a stabilizer. The performance of the stabilized soft clay soil from Dakahleia, Northern Delta region in Egypt was investigated and the conclusions were as follows:

1. The shear modulus, G, obtained after treatment with cement is improved significantly with increasing the cement percentage. Improving was notably observed in the soil stiffness and the maximum shear modulus. Similar response was also observed for using lime but the cement improving effect on G is greater at the same stabilized percentage.
2. The damping ratio, D, decreased by increasing stabilizer content.
3. At small to medium shear strain amplitudes, the decay in shear modulus value in case of lime stabilization is higher than the decay that occurs when cement is used as a stabilizer agent.
4. Threshold shear strain tends to decrease with treatment; at cement treatment, the apparent γ_{th} ranged from $2E^{-6}$ to $1E^{-5}$, while in case of lime treatment ranged from $1E^{-5}$ to $3E^{-5}$.
5. Based on results of maximum shear modulus (G_{max}); the appropriate cement content by weight of soil mass is more than 6%.; the appropriate lime content by weight of soil mass is not less than 8%.

References

Bergado, D.T., Anderson, L.R., Miura, N., Balasubramaniam, A.S.: Improvement of Soft Ground in Lowland and Other Environments. ASCE Press, New York (1996). 427 p.

Biarez, J., Gomes Correia, A., Liu, H., Taibi, S.: Stress-strain characteristics of soils interesting the serviceability of geotechnical structures. In: 2nd International Symposium on Pre-failure Deformation Characteristics of Geomaterials, Torino (1999, in press)

Cavallaro, A., Maugeri, M., Mazzarella, R.: Static and dynamic properties of Leighton Buzzard sand from laboratory tests. In: Proceedings of the 4th International Conference on Recent Advances in Geotechnical Earthquake Engineering and Soil Dynamics, San Diego (2001)

Chen, H., Wang, Q.: The behavior of organic matter in the process of soft soil stabilization using cement. Bull. Eng. Geol. Environ. **65**(4), 445–448 (2006)

FHWA: Ground Improvement Technical Summaries "Federal Highway", Administration Report No. FHWA-SA-98-086, Washington, D.C., USA (1998)

Ho, M., Chan, C.: Some mechanical properties of cement stabilized Malaysian soft clay. World Acad. Sci. Eng. Technol. **74**, 24–31 (2011)

Ingles, O.G.: Mechanics of clay stabilization with inorganic acids and alkali. Aust. J. Soil Res. **8**, 581–596 (1970)

Jaritngam, S., Swasdi, S.: Improvement for soft soil by soil-cement mixing. In: Proceedings of the Fourth International Conference on Soft Soil Engineering, Vancouver, Canada, pp. 637–640 (2006)

Kawasaki, T., Niina, A., Saitoh, S., Suzuki, Y., Honjyo, Y.: Deep mixing method using cement hardening agent. In: Proceedings of the 10th International Conference on Soil Mechanics and Foundation Engineering, Stockholm, vol. 1, pp. 721–724 (1981)

Lo Presti, D.C.F., Pallara, O., Lancellotta, R., Armandi, M., Maniscalco, R.: Monotonic and cyclic loading behavior of two sands at small strains. GTJ **16**(4), 409–424 (1993)

Mezni, N., Bouassida, M.: Contribution to the study of geotechnical characterization and behaviour of Tunis soft clay. In: Contemporary issues in Soil Mechanics. Proceedings of the 2nd GeoMEast International Congress on Sustainable Civil Infrastructure, Egypt, pp. 160–174 (2018)

Moh, Z.C.: Soil stabilization with cement and sodium additives. Proc. ASCE **88**(SM6), 81–105 (1962)

Punmia, B.C.: Soil Mechanics and Foundations, Thirteen edn. Laxmi Publications PVT. Ltd., New Delhi (1994)

Ramaswamy, S.D., Aziz, M.A., Kheok, S.C., Lee, S.L.: Cement stabilization of silty clay subgrades for road construction in Singapore. In: Proceedings of the English Regional Conference for Africa on Soil Mechanics and Foundation Engineering, Harare, pp. 413–420 (1984)

Rodriguez, A.R., Castillo, H.D., Sowers, G.F.: Soil Mechanics in Highway Engineering. Federal Republic of Germany, TransTech Publications (1988). 843 p.

Uddin, K., Balasubramaniam, A.S., Bergado, D.T.: Engineering behavior of cement-treated Bangkok soft clay. Geotech. Eng.-SEAGS **28**(1), 89–119 (1997)

Winterkorn, H.F., Pamukcu, S.: Soil stabilization and grouting. In: Fang, H.-Y., Van Reinhold, N. (eds.) Foundation Engineering Hand Book, 2nd edn., New York, pp. 317–378 (1991)

Review of Soil Improvement Techniques

Mohamed Mamdouh Abdel-Rahman(✉)

Housing and Building National Research Center, Giza, Egypt
mrahman.hbrc@gmail.com

Abstract. Foundation designers use deep foundation, when shallow foundation proved to have inadequate bearing capacity for the loading conditions, or the total and differential settlements are intolerable for the given structure. As an alternative approach, designers may consider one of the many soil improvement techniques to improve the geotechnical quality of the ground, i.e. bearing capacity and settlement characteristics. In choosing a soil improvement techniques, several factors should be considered including the suitability of the technique for the soil in place, the associate cost beside the durability and its performance over the year for the given load and geometry. These techniques may be used as a solution for problems in the ground for both existing and new constructions. However, there are several considerations, which engineers must examine before adopting any of these methods. This paper presents a review of the available soil improvement techniques, including their advantages, disadvantages/limitations, and provide applicability guide to practicing engineers for a given soil conditions and field applications. Design case study is presented proposing compaction grouting technique to overcome the collapse of bored piles. Selected field case studies from literature have been also presented using different soil improvement techniques explaining the improvement achieved from using each technique.

Keywords: Soil improvement techniques · Bearing capacity · Settlement · Case studies

1 Introduction

The main mission of the Geotechnical Engineer is to evaluate the soil conditions and characteristics then suggest the best design option that could fit with them. Shear strength, compressibility, and volume changes due to the hydrologic and environmental changes are the most important conditions that are always taken into consideration in any case study. Therefore, when the soil characteristics are good enough then installing shallow foundation would be a safe and economical choice. However, if the soil layer (such as soft or porous soil layers) is not suitable to support shallow foundation, then in that case, deep foundation is one of the solutions that could be applied by reaching a better soil layer such as sand or rock. Nevertheless, this solution is usually uneconomical, therefore, soil improvement techniques or soil stabilization methods which are based on changing the soil characteristics and conditions in order to increase its stability and

H. Shehata and M. Badr (Eds.): *Advancements in Geotechnical Engineering*, SUCI, pp. 175–199, 2021.
https://doi.org/10.1007/978-3-030-62908-3_14

make it suitable for foundation purposes are now being used and approved to be a great alternative (Cernica 1995).

The main purposes of using soil improvement techniques are: to increase the soil shear strength, increase the bearing capacity of the soil, reduce the hydraulic conductivity, reduce the soil settlement caused by external loads, reduce swelling and shrinkage of the soil mass in the active zone and increase safety factors in slope stability.

1.1 Soil Replacement

This is one of the oldest and simplest soil improvement techniques. It simply works by removing the soil layers that are not suitable for foundation purposes and replacing them with good, compact and suitable fill layer. The removed layer can be used later on as a fill layer after increasing its unit weight by compacting it more. This method is only able to be applied in case of soil layers subjected to collapse, very porous soil layer, very soft soil layer or if the soil layer contains organic or aggressive chemical substances. It should be mentioned that applying this method is difficult in case of ground water existence in the soil layer, even if we have pumps for removing the water, it will not solve the problem because it will still be hard to compact the soil layer because of the presence of water content in the soil layer.

1.2 Soil Pre-compression

It involves methods of ground improvement whereby the ground in question is subject to a surcharge load, usually in the form of a fill or embankment, causing a relatively higher rate of consolidation and thus reducing further settlement once the design load is applied. In its simplest application, which is common with fairly compressible soils, an embankment is placed on the ground in stages. Once the soft soil beneath has been compressed and

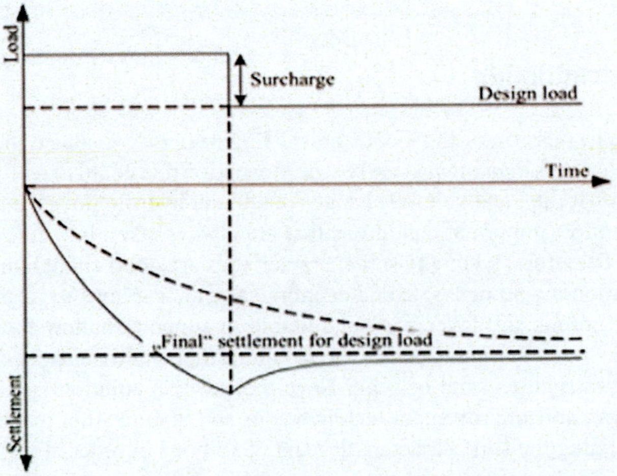

Fig. 1. Resulting settlement due to preloading (Stapelfeldt 2006)

settled past the pre-determined point for the design load, the embankment is then removed as shown in Fig. 1. Depending on the degree of over consolidation of the soil, secondary settlements after this process are greatly reduced and may be overlooked.

In situations with low permeability clay and silty soils, vertical drains are employed to accelerate the pre-compression process as shown in Fig. 2. Since clays are known to have higher permeability in the horizontal direction than the vertical, when vertical sand drains are introduced, a hydraulic gradient will allow for the excess water pressure, resulting from the preloading, to be released. This process can be viewed as a way of providing radial consolidation in addition to the vertical. Factors affecting the application of vertical drains technology include: the mechanical properties of the soil being investigated, and the type of vertical drains and method of installation (Stapelfeldt 2006).

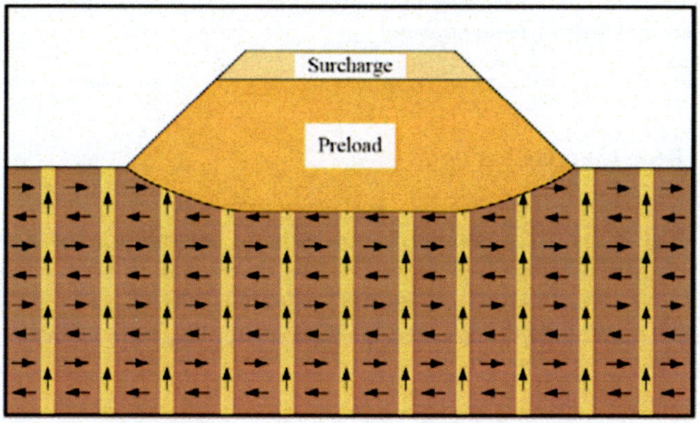

Fig. 2. Preloading with vertical drains (Stapelfeldt 2006)

In the case of much softer soils, the installation of vertical drains may disturb the soil layers reducing permeability. In addition, the soil could be unable to withstand the surcharge load itself. In this case, vacuum consolidation technique can be applied. The basic idea of vacuum consolidation is applying a vacuum suction into an isolated soil mass to reduce the atmosphere pressure and pore water pressure in the soil, resulting soil consolidation and effective stress enhance without changing the total stress. In contrast with the application of a preloading embankment, vacuum preloading will cause inward lateral displacement and consequently cause tension cracks along the periphery (Chai et al. 2014; Bhosle et al. 2015).

1.3 Soil Compaction

It is the method of soil stabilization which provides dense soil by applying stress or load on the soil helping to remove air voids and packed soil particles close to each other. In general there are four types of soil compaction: Vibration - Impact - Kneading – Pressure. For every compaction method, there are different compaction machines according to the purpose of compaction and the type of soil to be stabilized. Table 1 presents a comparison between different machines used in soil compaction (DeJong-Hughes et al. 2001).

Table 1. Comparison between Different Compaction Machines

Machine	Compaction Type	Suitable Soil			
		Gravel	Sand	Clay	Silt
Vibrating Sheepfoot Rammer	Impact			Excellent	Good
Vibrating Roller, Vibrating Plate Compactor	Vibration	Good	Excellent		
Grid roller, scraper	Pressure(with kneading)			Good	Very Good
Rubber-tired Roller, Grid Roller Loader	Kneading (with pressure)	Very God	Good	Good	Excellent

1.4 Deep Compaction

Comparing with the normal or surface compacting where it is valid only for depth less than 2.0 m, deep compaction gives more flexibility and reliability in improving soil characteristics in deep depths reaches up to 35.0 m. Deep compaction consists of two different methods to improve the soil characteristics, which are used to compact loose sandy layer with large thickness; namely: Dynamic Compaction, and Vibratory Compaction.

1.4.1 Dynamic Compaction

This technique is also known as dynamic consolidation or heavy tamping or TLM (Triple-Layer Model), which was found by the technician Louis Menard in the 1970s. It is inexpensive, fast and practical. It is based on applying high intensity impacts of free-falling heavy steel or concrete hammer on the surface of the soil layers. The dynamic waves caused by this hammer improve the characteristics of the soil layers in a considerable depth by making them denser and more compact. The effect of compaction on pore space is shown in Fig. 3.

The main soil layers targeted by this technique include: loose sandy silty soil layer in seismically active regions, soil layer that contains industrial remains and fragile rock, soil layer that is affected by liquefaction or soil layer that suffered from large settlement. However, in some cases, it causes some deterioration in the coarse soil layer which makes the surface compaction a necessity.

It is usually used to increase the bearing capacity of the soil layer and reduce the settlement at specific depths in order to open the opportunity to use shallow foundation. It should be mentioned that this technique is not suitable in some cases such as clayey soil layer where the hydraulics conductivity is very small which makes the consolidation process difficult and in case of projects underlain by landfills and peat. It should be taken into consideration that there is a small distance that should be kept between the

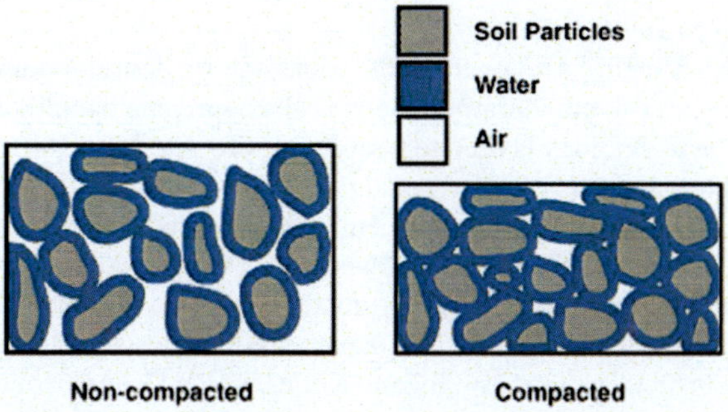

Fig. 3. Effects of compaction on pore space, Courtesy University of Minnesota

points where the steel or concrete hammer works and the adjacent buildings where the dynamic movement caused by this procedure may affect negatively any buildings nearby (McCarthy 2002).

1.4.2 In-Depth Vibratory Compaction

This technique is another economical and effective method of deep compaction and consists of two methods: a) Vibro Replacement in soft cohesive soils and inorganic fills; and b) Vibro Flotation for loose cohesionless soils. The main concept of this method is that soil could be compacted by means of a machine that would carry the source of vibration to the desired depth. The main equipment of this technique is an elongated heavy tubular steel vibrator that might be electrically or hydraulically driven, attached to some extensions and housed eccentric rotated discs with water jetting which washes the soil layer with coarse materials which are compacted in the next phase, as shown in Fig. 4.

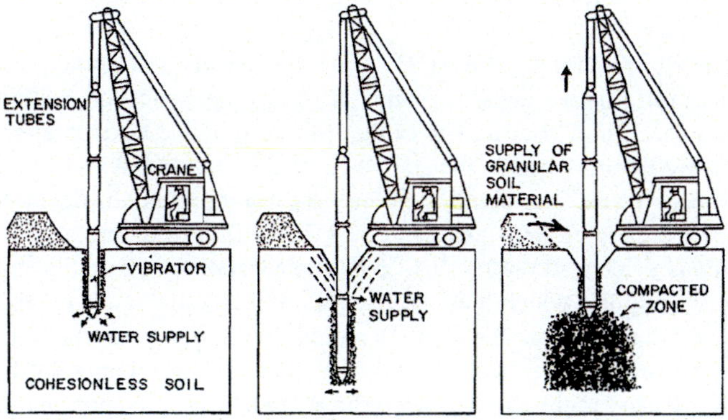

Fig. 4. Soil vibratory compaction method, *a) Chemical Grouting b) Cement Grouting*

1.4.2.1 Vibro Replacement

Vibro replacement is a method of constructing densely compacted stone/concrete columns using a depth vibrator to densify the aggregate backfill and surrounding granular soil. The technology is used to treat clays, silts and mixed stratified soils and improve their load bearing capacity and settlement characteristics.

One example of the vibro replacement technique is the Vibro Stone Columns (VSC) which involves the improvement of weak soils by the installation of densely compacted columns made from gravel or similar material with a vibrator. The displacement process reinforces all soils in the treatment zone and densifies surrounding granular soils. This technique is first developed by Johann Keller, Keller Group's founder. The advantages of this technique are: to reduce foundation settlement, increase bearing capacity, allowing reduction in footing size, increase stiffness, increase shear strength, reduce permeability, mitigate potential for liquefaction, permit shallow footing construction in treated fills, and very effective for sand compaction and land reclamation (Keller Group UK, n.d.).

Another example is the Rigid Inclusions, also known as Controlled Modulus Columns (CMC). It is a ground improvement method using high deformation modulus columns constructed through compressible soils to reduce settlement and increase bearing capacity. Ground improvement efficiency depends on the stiffness relationship between the soil and the columns. Load from the structure is distributed to the soil and columns via a load transfer platform or rigid foundation. The advantages of this technique is to allow the use of shallow foundation to support structure on compressible soils, increase the bearing capacity of weak soils, even for high loads, reduces settlements very efficiently, and it can be used in most construction sectors and applied for most types of structures and soil conditions (Keller Group UK, n.d.).

1.4.2.2 Vibro Flotation

This method was first found in Germany in 1930s and it's now used in USA and Europe. It is based on using a Vibro-Flot, which is a kind of vibrating unit. This method is useful in case of coarse and medium sand layer under the water table. Whereas it is limited when soil particles get finer or some silty soil particles exist because the area compacted gets smaller.

The vibroflotation process (VF) is used to densify soil formations which do not have an optimum relative density, mainly naturally deposited or backfilled granular soils, such as sands and gravels. Under the influence of the horizontal vibrations generated by the oscillating vibrator, the soil particles are rearranged and adopt a denser packing. After reaching the final depth, supported by water jetting, the depth vibrator is gradually retracted creating a densified zone of 2.0 to 4.0 m in diameter. The reduction in pore volume is evidenced on the surface by the formation of a settlement crater around the compaction point which must be backfilled with suitable coarse material. This allows the material to densify and creates a column with improved engineering characteristics, including an increase in strength and a reduction in compressibility. It increases the relative density of the soil, and accordingly reduces settlements and improves resistance to liquefaction (BAUER, n.d.).

1.5 Grouting

It is the kind of a soil stabilizing technique in which special form of a stabilizing agent injected under pressure into the soil. It is much more expensive as compared to other soil stabilization methods. Usually, the admixture of cement and water (with or without sand) is used as a stabilization agent. The pressure forces the admixture to fill the voids of the soil and make it stable to have solid soil which has good bearing capacity. It assists the soil to reduce permeability through filling the voids and cracks of the soil deposit. Most often we prefer grouting system when the permeability (k) of the soil is beyond 10 – 5 m/s (Hayward Baker, Grouting, n.d.). Figure 5 (a) and (b) show chemical and cement grouting techniques, respectively. Some of the commonly used forms of grouting are:

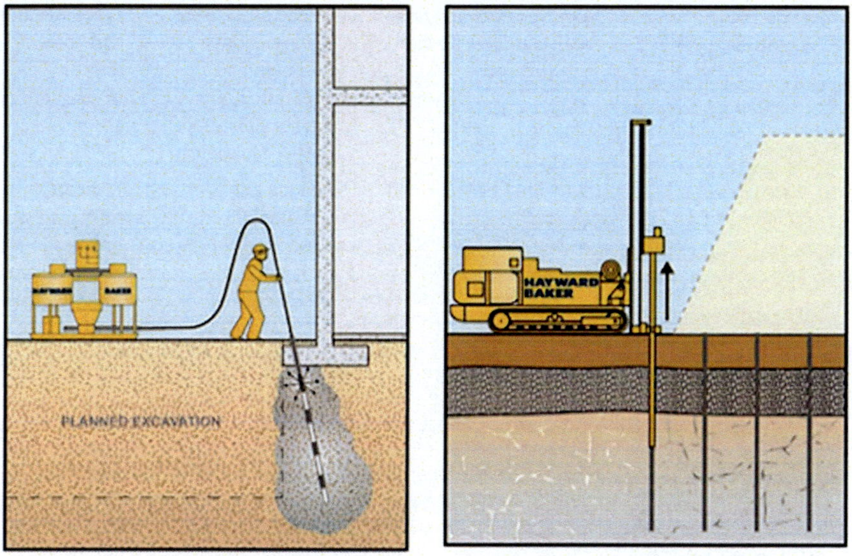

a) Chemical Grouting **b) Cement Grouting**

Fig. 5. Grouting soil improvement techniques, courtesy hayward baker

1- Chemical grouting
2- Cement grouting (slurry grouting)
3- Compaction grouting (low mobility grouting)
4- Jet grouting
5- Fracture grouting.

They all have their own advantages and disadvantages; Table 2 presents a comparison between these methods according to the type of soil, purpose, and application (Hayward Baker, n.d.).

Table 2. Comparison between Different Grouting Techniques

Grouting technique	Soil type	Purpose	Application
Chemical Grouting	Granular sand and gravel soil Best suited for sand with low fine particles	Improve strength and stiffness, and eventually reduce permeability	Underpin existing substructures, create a water cut-off barrier, Stabilize tunnel soils etc.
Cement Grouting (High Mobility grouting)	Granular soil and rock	Increase strength and stiffness	Reduce the flow of water through rock deposit underneath dams etc.
Compaction grouting	Loose, coarse grained soils and densify fine grained soils	Enrich soil bearing capacity, reduce settlement and liquefaction, also to steadies subsurface voids or sinkholes	Treat standing sinkholes or to diminish its potential in sinkhole prone areas.
Jet Grouting	Any kind of loose soil (soft clay, sand and gravel)	Enhance the mechanical appearances of the soil	Projects with restricted space, underpinning, retaining walls, impervious cut-off wall etc.

1.6 Soil Stabilization Using Chemical Admixtures

Geotechnical engineers have been making use of soil stabilization techniques to improve gradation, reduce plasticity index and swelling potential, and increase strength and durability for a long time. Various applications of soil stabilization techniques exist and have been evaluated; however most common is the addition of various chemical compounds to the soil depending on the qualities to be achieved and the economic results among other factors.

Most commonly additives employed are Portland Cement admixtures which improve soil strength and durability as well as manage the erosion within inorganic soils. Numerous past studies have shown a strong correlation between cement content and unconfined compressive strength. Furthermore, it has been shown that the increase in cohesion and friction angles by cement admixtures is improved with an increase in curing time. In addition, cement treatment provides moisture resistance. Cement stabilization is most common with weak expansive soils with an organic content less than 2% and pH higher than 5.3. In spite of it is being widely used around the world, for reasons involving budgets and scheduling, many construction projects might opt for alternatives (Kowalski and Starry 2007).

Lime is a readily available alternative to cement admixtures. Lime admixtures typically decrease plasticity and improve consistency of fine grained soils. Strength improvement, however, is not always an outcome. Lime's contribution to strength improvement comes by way of flocculation of soil particles and this is most evident in soils with large surface area. These and other aspects of lime admixtures depend on a wide range of properties including the soil and lime types as well as curing conditions. A number of methods have been developed to determine the optimum percentages of lime to be added to the soil (Negi et al. 2013; and Ashraf et al. 2018).

Overall, lime is favored as an admixture option because in addition of being a relatively simple and requires less labor intensive process, it is also significantly inexpensive. This makes it conducive for small scale applications. Since compaction and gradation are greatly improved, lime is most effective for soils exposed to frequent expansion and shrinkage.

Similar in action to lime is another alternative admixture, Fly Ash, produced in two types: type C and type F. It acts to increase dry density of the soil as well as to reduce optimum moisture content. Class C fly ash can be used as a standalone material because of its self cementitious properties. Class F fly ash can be used in soil stabilization applications with the addition of a cementitious agent (lime, lime kiln dust, CKD, or cement).

Flay Ash can be used to dry wet soils and provide an initial strength gain by reducing the amount of clay sized particles. It can be used where the soil material exhibits high volumetric changes. It is a suitable choice for the purpose of improving gradation, enhance strength properties, stabilize embankments, and to control shrink swell properties of expansive soils (Nalbantoglu 2004).

Among the benefits of using fly ash is that it brings the soil to a relatively more granular consistency effectively decreasing water absorption potential in wet soft soils. Fly ash may also have varying chemical properties depending on their type. Some forms may have more abundant free lime allowing pozzolanic and cementitious reactions (Ramaji 2012).

The versatility of fly ash which may be used in combination with other admixtures makes it a viable option taking into account the physical properties (bulk density, permeability, gradation) of the soil to be stabilized.

1.7 Geotextiles

Geotextiles have been used to provide a tensile strength element to a soil mass when sited underneath foundation. This tensile reinforcement helps to reduce stresses and strains within the soil mass or foundation and empowering the substructure to resist differential settlements and lateral distribution or movements. Geotextiles reinforcement in soil foundation assist to: empower the safety, raise the superstructure height, and resist stresses and strains.

1.7.1 Use of Geotextiles

a) Weak Foundations: geotextiles reinforcement have been used to stabilize week or soft foundation soils, i.e. deep soft soil, saturated clay, silt or organic soil deposit.

These kinds of soils spread laterally when they are under horizontal earth pressure and surcharge or under the influence of high loads. Weak soils are not able to resist these stresses, so geotextile reinforcement is used, as shown in Fig. 6, to provide soil reinforcement and stabilization (FEMA, Geotextiles in Embankment Dams 2008).

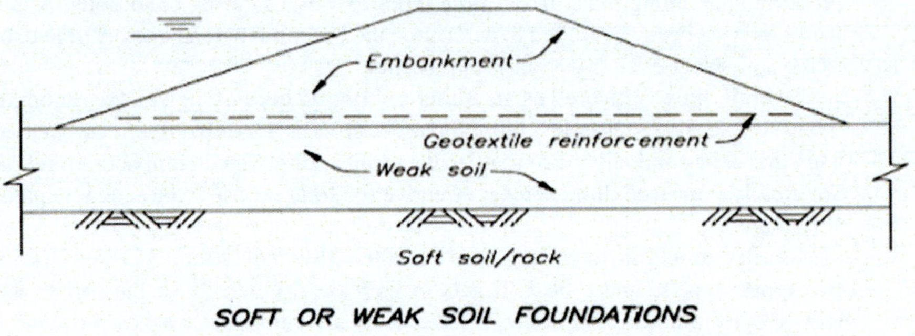

Fig. 6. Geotextiles in embankment dams – single layer (FEMA 2008)

b) Voided Foundations: voids in foundation are caused by old stream beds, sinkholes, soluble bedrocks, etc. In these kinds of foundations, geotextiles have been used to deliver tensile reinforcement in both directions by placing geotextiles layers over these voided foundations, as shown in Fig. 7 (FEMA, Geotextiles in Embankment Dams 2008).

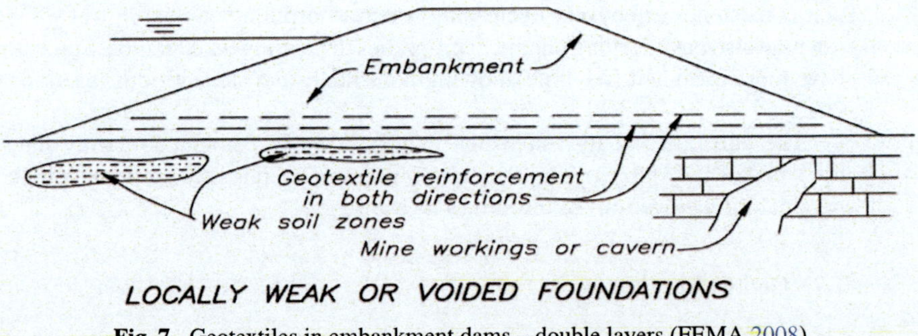

Fig. 7. Geotextiles in embankment dams – double layers (FEMA 2008)

1.8 Ground Freezing

This technique cannot be applied in expansive soils, coarse gravel or boulder soils; however it gives better results in soft soil than rock formations regardless of the size, shape and depth of excavation. The purpose of applying this technique is to:

a) Reduce the need for dewatering or compressed air.
b) Provide temporary support or structural underpinning for an excavation.

c) Reduce probability of having subsidence in adjacent ground during excavation especially in case of soil particles immigration during freezing process.
d) Prevent groundwater in an excavated area to flow; therefore ground water table should be taken into consideration for the whole excavated area and the adjacent one in order to eliminate the opportunity of having ground water precipitation which causes failure in the freezing procedure.

Ground Freezing technique could be done rapidly or slowly, however the fast way is much more costly than the slow one, where the freezing plants take 40 to 60% of the main budget. Since, this technique is based on freezing, the ground that means thermal energy exchange occurs during the freezing process. The amount of energy required to freeze the soil particles is about 9 to 12 times the percentage of existing groundwater. Therefore, the energy required to freeze fine grained silt and clayey soils is higher than the energy needed to freeze coarse grained soils.

In order to have an efficient freezing system, detailed investigation must be done to the whole subsurface formation that is going to be frozen in order to estimate accurately the soil characteristics, groundwater table source, site location, and the real temperature of the ground. This could be done by doing several boreholes that cover all the project area with sufficient depth and keep some samples as reference for later revision if needed (Sego et al. 1994; Andersland and Ladanyi 2004).

1.9 Blast Densification

It is also known as Explosive Compaction (EC), it is useful to increase the bearing capacity, decrease the susceptibility to both static and seismic induced liquefaction and decrease the quantities of seepage in the order to reach the point of densification in the loose saturated sandy soils. It is based on making a borehole with 150 mm in diameter by jetting or drilling depending on the soil characteristics, then installs explosives at certain depths in order to detonate them in an efficient systematical pattern.

Moreover, it is considered one of the most economical methods in case of projects with huge amount of excavations, also it gives better effective results in case of soils that consist of two layers, and a loose sand layer overlies a dense sand layer. Installing charges is usually done between 1/3 to 2/3 within the depth of loose layer in order to protect the dense layer from weakening. It should be mentioned that the SPT tests need several weeks to reflect the improvement done in the soil. The efficiency of this technique predominantly depends on the soil profile, grain size distribution, initial status, and the intensity of energy applied to the soil. It has been observed that it can be an effective method to improve the density, stability, and resistance of the target soils (Shakeran et al. 2016).

1.10 Electro Osmotic and Electro Kinetic Stabilization

This technique is based on an interaction of an electric field, where this interaction happens from using a direct current DC voltage that could be applied on the soil layer by a positive terminal (electrode poles) whereas the negative terminal (cathode poles) attract the soil pore water towards it. The main purposes of using this technique include:

consolidation and strengthening of soft saturated clayey soils, strengthening silty clays and soft sensitive clays, increasing the slope stability, increasing the reinforcement of the steel piles in clayey soils and improvement of the soft clayey soils. It should be mentioned that adding drainage to the technique at the cathode side only makes electro-osmosis induces the consolidation procedure, which results in lower soil water content, higher shear strength and lower compressibility. Moreover, physical and chemical properties of the soil are altered and shear strength is increased by using electro-osmotic process associated with electrochemical reactions (Chien et al. 2012).

2 Comparative Study

A comparative study between several soil improvement techniques using mechanical or chemical methods in relation to the suitability of soil conditions, advantages, disadvantages/limitations and field applications is summarized in Table 3 is presented as a guide to practicing engineers for selecting the proper soil improvement technique.

3 Design Case Study

3.1 Problem Definition

The project is a high-rise building located near Cairo, Egypt and its foundation is designed to rest on 200 post-grouted cast- in-situ bored piles with a diameter 1.20 m and an axial load of 1270 ton. The piles length below the bottom level of the raft varies from 30.00 m to 33.00 m. Piles shafts were bored dry using a 6 m long temporary casing during the boring process. Non-working piles were tested prior to commencing the work. Shaft instability problems were encountered in 13 bored shafts and collapse of the boring was reported at different depths for these 13 piles. Table 4 summarizes the encountered problems for each of the 13 pile shafts and the current status of the pile based on the action taken by the piling contractor. It was reported that after soil collapse was stabilized via casting cement mortar, the drilling work was resumed and the defective piles were completed except for two piles where the Kelly was buried in the collapsed shaft (No. 1 and 2).

3.2 Subsurface Conditions

Based on the conditions encountered in the boreholes, the native deposits along at the site generally consisted of a layer of sand deposits which overlies alternating thick layers of claystone (hard clay) and cemented sand (to silty sand). Another layer of sand underlies these layers. The soil stratification at the project's site can be divided into four main layers as follows:

3.3 Layer no. 1: Sand

A surface layer extends from the current ground level up to depth about (3.0 m–29.0 m). The layer consists mainly of sand, together with different percentages of silt and gravel. A thin layer of fill occasionally overlies this sand layer. This Sand layer is usually yellowish brown, medium to coarse grained, and very dense. The layer is sometimes intercalated with interchangeable layers consisting either of sandstone or cemented sand.

Table 3. Comparison between Different Soil Improvement Techniques

	Technique	Soil Conditions		Advantages	Disadvantages /Limitations	Field Applications
		Suitable	Not Suitable			
Mechanical	Soil Replacement	- Soil layer subjected to collapse - Very porous soil layer - Soft soil layer - Soil layer contains organic material - Soil layer contains aggressive chemical material	- Soils with ground water existence, because compacting works is hardly to be done	- Simple - Less delay for construction - Economical	- Case of ground water existence in the soil layer	- Low-rise buildings - Shallow foundations (isolated – raft)
	Soil Pre-compression	- Compressible soils - Low permeability clay and silty soils with the application of vertical drains	- Soft soil layers which unable to withstand surcharge	- Reduce post-construction settlement - Reduce secondary compression - Soil densification - Improve bearing capacity - Almost free of noise and vibrations	- Insufficient time to achieve required compression or pore-pressure dissipation - Excessive lateral deformation of the pre-compression fill - Inadequate factor of safety against shear failure	- Buildings - Embankments - Harbors - Highways - Large Tanks
	Vacuum-preloading	- Soft soil layers which unable to withstand surcharge	- Expansive soils	- Does not require heavy machinery - Instantaneous removal of the surcharge - Avoidance of failures due to reduction of shearing stresses under embankment - Increase gradient and consolidation rate	- Applications in expansive soils	- Buildings - Embankments - Harbors - Highways - Large Tanks - Dams

(*continued*)

Table 3. *(continued)*

	Technique	Soil Conditions		Advantages	Disadvantages /Limitations	Field Applications
		Suitable	Not Suitable			
	Dynamic compaction	- Loose sandy silty soil layer in seismic active regions - Soil layer contains industrial remains - Fragile rock or soil layer that is affected by liquefaction	- Clayey soil with low hydraulics conductivity - Sites underlain by landfills and peat - Expansive or organic soils	- Increase bearing capacity - Reduce settlement - Mitigate liquefaction - Avoid major below-ground obstruction removal - Capable of compacting variety of weak soils - Sustainable using limited material import	- Soil layer that suffered from large settlements - Vibrations could cause disturbance and possible damage of surrounding structures - Caution should be used on soil profiles including cohesive soils	- Embankments - Harbors - Highways
Mechanical	Vibro replacement	- Soil layers subjected to collapse - Porous soil layers - Soil layer contains organic material - Soil layer contains aggressive Chemical material	- Not efficient in case of fine grained soils such as clayey or silty Soil - Partially saturated soils	- Can be applied under ground water - Reduce settlement - Increase bearing capacity, stiffness and shear strength - Reduce permeability - Mitigate liquefaction - Increased slope stability - Very effective for sand compaction and land	- Vibration during installation could harm the neighboring buildings - Difficult to penetrate stiff overlayers	- Embankments - Harbors - Highways
	Vibro flotation	- Coarse and medium sand layer under ground water table	- Fine soil particles or silty soil particles - Coarse and fine gravel	- Reduce differential settlements - Fast, easy, cost effective - Done around existing structures - Improve soil using its own characteristics - Reduce liquefaction due to earthquakes	- Densification generally cannot be achieved when the granular soil contains more than 12 to 15% silt or more than 2% clay	- Harbors - Underpinning

(continued)

Table 3. (*continued*)

	Technique	Soil Conditions		Advantages	Disadvantages /Limitations	Field Applications
		Suitable	Not Suitable			
	Ground freezing	- Soft soil rather than rock formation regardless size, shape and depth of excavation.	- Expansive soils, coarse gravel, and boulder soils	- An extremely versatile method for temporary ground improvement or cutoff - Applicable to difficult ground conditions	- Soil has to be near saturation or completely saturated - Highly energy intensive process - Requires plenty of monitoring such as: temperatures, deflections of nearby structures, heaving and settlement at ground surface - Expensive technique	- Open excavation support and groundwater control - Environmental applications for GW contaminants - Tunneling
	Blast Densification	- Loose saturated sandy soils, effective in case of soil consist of two layers a loose sand overlies a dense sand	- Soils with relative density, DR $\geq$ 50–60%. - Unsaturated undrained soils, Clay > 5–10%., Silt > 70–80%.	- Economical in projects with huge amount of excavations	- May cause damages to nearby existing structures - Hazardous - No soil improvement near surface	- Prevent liquefaction and flow in saturated, loose sandy soils in earthquakes. - Removing and replacing soft soil layer with crushed stones
Chemical	Compaction grouting	- Most effective for fine grained soils with sufficient permeability - Variety of soil and fills - Suitable for rubble fills, poorly placed fills, loosened or collapsible soils, soluble rocks and liquefiable soils	- Expansive soil - Soils with high organic content	- Reach to a greater soil depth without the help of other methods - No environmental impact - Densify the soil and thereby increase the bearing capacity - Reduce settlement	- Improper control of the pressure applied may damage the soil structure	- Often selected for treatment beneath existing structures - Mine stabilization - Underpinning foundation problems - Tunneling

(*continued*)

Table 3. (*continued*)

	Technique	Soil Conditions		Advantages	Disadvantages /Limitations	Field Applications
		Suitable	Not Suitable			
	Stabilization using Portland Cement	- Weak expansive soils - Soil with organic content < 2% and pH > 5.3	- Soil layer with high ground water - Clay soil with sulfate	- Less expensive than soil replacement - Improve soil bearing capacity	- Can be ineffective in heavy freeze/thaw areas - Can be in effective in high ground water areas	- Highways - Structural fills - Embankments
	Stabilization using Hydrated Lime	- Soils exposed to frequent expansion and shrinkage - Subgrade soil containing clay	- Soil layer with high ground water	- Less expensive than soil replacement - Improve soil bearing capacity	- Strength is not always an outcome with relatively small surface area of soils	- Highways - Structural fills - Embankments
Chemical	Stabilization using Fly Ash	- Soil which exhibits drastic volumetric changes - Wet soft soils with high water absorption potential	- soils containing more than 10% sulfates have been proven to swell excessively in some applications - Organic soils are difficult to stabilize using fly ash	- Eliminates need for expensive borrow materials - Expedites construction - Can reduce or eliminate the need for more expensive natural aggregates in the pavement cross-section - Cost savings	- Poor quality fly ash may have negative effects on strength - Caution must be directed to moisture content and hydration reaction - It may be subject to local environmental requirements pertaining to leaching and potential interaction with ground water and adjacent water courses. - Fly ash with a sulfate content greater than 10% may cause soils to expand more	- Highways - Structural fills - Embankments

(*continued*)

Table 3. (*continued*)

	Technique	Soil Conditions		Advantages	Disadvantages /Limitations	Field Applications
		Suitable	Not Suitable			
	Electro Osmotic & Electro Kinetic Stabilization	- Soft saturated clayey soils, silty clayey, soft and sensitive clays	- Clay with interbedded sand because the injected chemical solutions might escape from sand layer - Highly conductive soils	- Increase slope stability - Effective improvement for soft clayey soils - Cost effective - The site conditions can be both in-situ and ex- situ - Simple in operation - No noise pollution - Other site activities are not affected - Treatment time period is short	- Heat generation in excessive amount during treatment - Cracking of electrodes due to heat generation - Electrochemical reaction results in chlorine gas which is harmful - High variation of pH value resulting to degradation of anode or cathode electrodes - Improper usage of Chemical agent to affect the whole system	- Embankments - Open excavations - Slope stability - Strengthening of backfill

Table 4. Summary of the encountered pile shaft instability

No.	Description of encountered instability	Action Taken	Pile Status
1	Upon drilling to a depth of 33 m, the drilled hole collapsed and Kelly was buried. Only 13 m of the drilled hole remained	Two Kelly sections were pulled out and one is still buried along with bucket	Stalled
2	Upon drilling to a depth of 31 m, the drilled hole collapsed and Kelly was buried. Only 8 m of the drilled hole remained	Two Kelly sections were pulled out and one is still buried along with bucket	Stalled
3	Upon drilling to a depth of 15 m, the drilled hole collapsed. Only 5 m of the drilled hole remained	The remaining hole depth was cast cement mortar and casing was pulled out	Completed
4	Upon drilling to a depth of 14 m, the drilled hole collapsed. Ground settlement was noticed around the casing at the platform level	Drilling resumed	Completed
5	Upon drilling to a depth of 5 m using a 5 m casing, the drilled hole collapsed	12 m casing was used and drilling resumed	Completed
6	Upon drilling to a depth of 13 m, the drilled hole collapsed. Only 8 m of the drilled hole remained	Drilling resumed	Completed
7	Upon drilling to a depth of 12 m, the drilled hole collapsed	The remaining hole depth was cast cement mortar and casing was pulled out	Completed
8	2.5 m from platform level collapsed prior to casing installation	Drilling resumed	Completed
9	Upon drilling to a depth of 13 m, the drilled hole collapsed. Only 8 m of the drilled hole remained	The remaining hole depth was cast 10 m^3 of cement mortar and casing was pulled out	Completed
10	Upon drilling to a depth of 3 m, the drilled hole collapsed. Only 25.5 m of the drilled hole remained	The remaining hole depth was cast 30 m^3 of cement mortar and casing was pulled out	Completed
11	Upon drilling to a depth of 9 m using a 5 m casing, the drilled hole collapsed and only 4 m of the drilled hole remained	12 m casing was used and drilling resumed	Completed
12	Upon drilling to a depth of 13 m, the drilled hole collapsed. Only 6 m of the drilled hole remained	The remaining hole depth was cast cement mortar and casing was pulled out	Completed

(continued)

Table 4. *(continued)*

No.	Description of encountered instability	Action Taken	Pile Status
13	Upon drilling to a depth of 30 m, the drilled hole collapsed. Only 11 m of the drilled hole remained. Ground settlement was noticed around the casing at the platform level	The remaining hole depth was cast cement mortar and casing was pulled out	Completed

3.4 Layer no. 2: Cemented Sand

A layer consisting mainly of cemented sand is interchangeable with a claystone layer up to depths about (10.0 m to below the tip of the pile). This layer is usually yellowish brown, friable to highly cemented, medium to coarse grained with traces of gravel and fines. The layer is sometimes intercalated with sandstone.

3.5 Layer no. 3: Claystone

A claystone layer underlies the surface sand layer up to depth about (29.0 m to below the tip of the pile). The layer consists mainly of claystone and is usually grey to brown, very stiff to hard. The layer is sometimes intercalated with cemented sand.

3.6 Layer no. 4: Lower Sand

Another layer of sand underlies the cemented sand / sandstone layer. The sand layer appeared in some boreholes at a depth of 29 m and extended to below the tip of the pile.

3.7 Potential Causes for the Collapse of Bored Shafts

According to the information provided by the contractor, all defective pile shafts were bored dry and no drilling fluids were used. The following could be the causes for the instability that occurred in the 13 bored shafts:

1- Boring non-cohesive soils (layers no. 1 and no. 4) without using a drilling fluid or casing.
2- Relatively long wait times between boring and casting of concrete might also cause instability of the bored shafts
3- Boring induced vibrations might have triggered the collapse of slightly cemented sand which may lead to progressive unraveling of non-cohesive soil into the bored cavity.

3.8 Proposed Solutions

The encountered soil collapse due to drilling has caused significant disturbance and loosening to the soil surrounding the pile shafts. The estimated shaft resistance of the pile might not be valid for the defective pile shafts. For the piles that have already been completed, cement mortar cast in the shaft void may result in the post-grouting process not working as it is supposed to. The following subsections propose mitigation measures for advancing new piles and suggest remedial measures for the soil surrounding the defective shafts.

a) Boring using Bentonite Slurry

It is recommended to resume the remaining boring work using a drilling fluid composed of bentonite and water. Additives can be used to enhance the properties of the bentonite slurry as necessitated by the boring conditions. The plasticity index of bentonite should be greater than 250%. The contractor should start with piles adjacent to the defective piles in order to confirm the success of the modified boring method (with bentonite slurry).

b) Proposed Methodology for Remediation

Based on the provided information and the site observations, it is expected that the soil surrounding the pile has experienced non-uniform displacement and loosening. It is also expected that cavities might have developed in the surrounding soil along the defective shafts. This may result in tangible reduction in the geotechnical capacity of the defective piles as well as the adjacent piles. The following subsections outline the recommended methodology for taking the remediation decisions.

b.1) Testing of Defective Piles

Pile load tests shall be performed on one or more of the 13 defective piles to assess the actual capacity and decide whether additional measures are required. Selection of the test pile(s) should be based on the construction records (e.g. bored pile reports) of the piles after all information is completed. The project consultant is to assess whether the reduction (if any) in the capacity of the defective piles is acceptable or not.

b.2) Recommended Remedial Measures

If, according to the tests, the reduction in pile capacity is found to be significant, remedial measures should be taken to compensate for the reduced capacity. The following presents the recommended remedial measures:

b.2.1) Additional Piles

Advancing additional piles around the defective piles is recommended to compensate for the drop in the capacity of the defective piles. The capacity and characteristics of the additional piles may require further assessment by the project designer.

b.2.2) Compaction Grouting

In case advancing additional piles is not possible, affected soil surrounding the additional piles should be improved. Two objectives should be set for the improvement technique to be used, soil displacement for soil densification and filling of the generated cavities. Injecting cement mortar or high slump grout to fill the cavities could have adverse effects such as softening the soil adjacent to the cavities and the development of planes of discontinuities that could initiate failure upon loading.

One of the techniques that may fulfill both objectives is "Compaction Grouting". This technique involves injecting the soil with low slump (less than 25 mm) grout under pressure. The injected grout does not permeate into soil but rather forms grout bulbs that displace and compact the surrounding soil. A column of bulbs can be formed successively from bottom up as shown in Fig. 8.

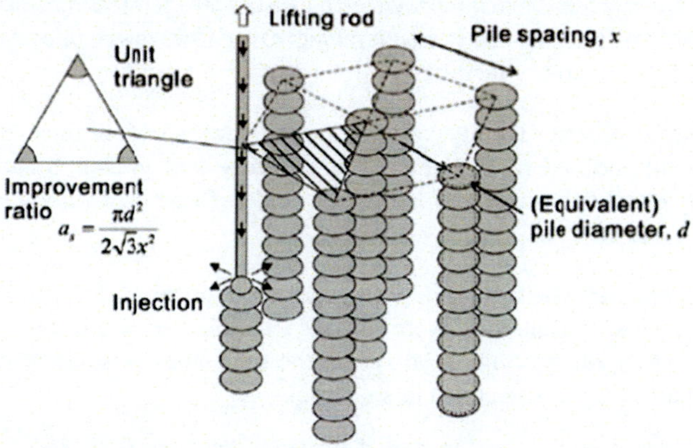

Fig. 8. Compaction grouting construction sequence (bottom-up) in triangular pattern (after Nishimura et al. 2011)

The grout material is normally soil cement mixture with sufficient silty fines and sand to provide plasticity and develop internal friction. Grout expands into the soil under pressure until refusal is reached. The size of the grout bulb depends on the stiffness of the surrounding soil, however, reaching a diameter of 1.0 m or larger is common in loose soils. Such dimensions are reported for continuous soil medium with no voids or cavities. In case of existing cavities, grout will fill any void before compacting the adjacent soil (Moseley and Kirsch 1993 and ASCE 2007).

4 Field Case Studies

Case (1): Missouri River in the City of Maryland (Khaldoun and Brandon 2001)**:** 5 stories office building using Dynamic Compaction.

Subsoil Condition: 15 borings were done in the site. A thick sequence of flood plain alluvium that included generally low strength and compressible clay, silt and sand (loose to medium). The geotechnical assessment of the site indicated the potential liquefaction of the loose sand at depth. Liquefaction analysis shows that number of blows at 4.5 m depth is lower than the required blow to prevent liquefaction at site of 4.5 m depth.

Improvement: Atotal of 29 boring were done and SPTs were performed after dynamic compaction. A significant improvement was observed to include:

1. Reduced liquefaction hazards.
2. Used spread footing instead of deep foundation.
3. Construction cost saved by $ 120000.
4. On time completion.
5. Owner's investment security.

<u>**Case (2)**</u>**: Lake Placid NY** (Gruner et al. 2001)**, using Micro-piles**: place to host two winter Olympic, had a major project to construct refrigerated truck at mount Vanhoevenberg for luge, bobsled and skeleton racing. Over 1300 micro piles were used to support the track and ancillary facilities.

Subsoil Condition: Highly variable ranging from uncontrolled granular fill, dense glacial fill and bed rock. Natural fills contains well graded granular soil containing varying quantities of silt and numerous bolder. Depth of bedrock is variable. Ground Water Table is 2.0 m.

Improvement: Micro-Piles (1300) were recommended for soil improvement. Two types of piles (soil and rock) were constructed. Soil piles length covers frost zone (2.4 m) and bond length in soil (4.5 m) while the rock piles was varied in length but had a minimum of 1.5 m embedded in solid rock.

Observed performance: 6 single piles (3 in soil and 3 in rock) and 9 group piles (6 in soils and 3 in rocks) were selected to test for

1. Tensile pile loaded test
2. Compression test
3. Lateral load test

Micro-piles proved to be a simple and cost effective solution to a complex foundation design for a combination of luge/bobsled racing track.

<u>**Case (3)**</u>**: Carrefour Shopping Center** (Martin et al. 2002), **using Jet Grouting** located in the city of Izmit, Turkey, approximately 8 km from earthquake epicenter. The total area is 55000 m^2 will house 2 main facilities and a parking lot.

Subsoil Condition: 17 SPTs and 15 CPTs were done: 0–2.5 m irregularly sandy gravel fill. 2.5–6 m medium clay. Below the clay fine to silty sand of loose to medium dense condition of 2–4 m thickness across the site. GWT is within 2 m.

Improvement: Surcharge fill were used to improved soft clay and Jet grout columns were used to improve bearing support in the clay as well as to prevent liquefaction. After earthquake (7.4 M) no structural damage was observed. No noticeable settlement and ground damaged were observed anywhere in the site.

Case (4): **The reconstruction of Interstate 15 in Salt Lake City, Utah** (Saye et al. 2001), **using Lime and Cement Columns**, required extensive ground stabilization measures to support the new roadway embankment. Lime cement columns were a part of the ground stabilization program used to support a 12.5 m wall and embankment between two bridges.

Subsoil Condition: The alignment is underlain by compressible soil extended to depth of about 20 m below the original ground surface consist of a series of clay strata that are separated by deposit of silt, sand and clay.

Improvement: The treated ground was significantly stiffer and the stability of the lime cement treated zone was better than expected. Zone of lime cement treatment settled quickly with complete consolidation occurring within 40 days of loading.

Case (5): **Part of Amtrak's Northeast corridor High-speed rail electrification project** (Benedict et al. 2001), **using Compaction Grouting**, it was necessary to extend the track 4 siding near Attleboro, MA to provide adequate sight distance for new signal. A remedial design was required to mitigate settlement and providing adequate slope stability considering the existing subsoil condition.

Subsoil Condition: Laboratory tests of peat indicated that it was normally consolidated with natural water content about 400%, saturated unit weight 11.9KN/cum, natural void ratio 11.7 and a compression index of 6.5.

Improvement: It was decided to use Compaction Grout column. 289 columns of 305 mm diameter with average 6 m length were installed in the site to improve the condition. In order to get the additional shear strength and thereby increased factor of safety against global stability failure, each column had reinforced with 1 #8 steel bars at the center of column. The unique application of this technology not only improved the stability of railroad embankment but also mitigate settlement concern resulting from soft and thick peat.

5 Conclusions

In the literature, several soil improvement techniques can be found, each has its advantages and limitations. This paper can be used to help as a guide to practicing engineer to select the appropriate technique, which is suitable for a given soil condition and field application.

Comparison of various soil improvement methods is difficult as each method has its own characteristic. However, selection of appropriate method is important to get the desired performance. In order to choose an appropriate method some factors could be taken into consideration including: stability requirement, allowable total settlement, rate of settlement, seepage criteria, durability and maintenance requirement, volume of soil to be treated, soil type, initial, depth of GWT, availability of materials, equipment and skills, cost, time available and environmental factors due to construction.

Soil improvement techniques are suitable in improving the soil condition to allow using shallow foundation instead of deep foundation. It is also suitable to reduce the down drag force in case of using piles, to eliminate the thickness of pavement layer in case of road design or to reduce the water precipitation in case of channels and dams. Each case has its specific conditions and circumstances that should be taken into consideration when a soil improvement technique is being selected.

Comparative study was presented between different types of soil improvement techniques in relation to applicability of soil conditions, advantages, disadvantages/limitations and field applications. Design case study was presented proposing compaction grouting technique to overcome collapse of bored piles. Selected field case studies from literature have been presented using different soil improvement techniques explaining the achieved improvement for each case.

References

Andersland, O.B., Ladanyi, B.: Frozen Ground Engineering. ASCE and Wiley, New York (2004)

ASCE: Compaction grouting consensus guide DRAFT. Compaction Grouting Guide Committee of the Geo-Institute of the ASCE. Reston (VA): ASCE/GI 53–10 (2007)

Ashraf, M.A., Hossen, M.A., Ali, M.A., Chakaraborty: Stabilization of soil by mixing with different percentages of lime. In: Proceedings of the 4th International Conference on Civil Engineering for Sustainable Development ICCESD 2018, Kuet, Khulna, Bangladesh (2018)

Bauer Group (n.d.). "Ground Improvement". https://www.bauer.de/bauer_group/index.html

Benedict, C., Haider, T., Byle, M.: Compaction grout columns for track support, Foundation and Ground Improvement, Geotechnical Special Publication No. 113, ASCE, Blacksburg VA (2001)

Bhosle, S.P., Deshmukh, V.B., Pawa, S.D.: Ground improvement of soft clay by vacuum preloading – a review. In: Proceedings of the 50th INDIAN Geotechnical Conference, 17–19 December 2015, Pune, Maharashtra, India (2015)

Cernica, J.N.: Geotechnical Engineering: Soil Mechanics. Wiley, New York (1995)

Chai, J., Liu, M.D., Carter, J.P.: Methods of vacuum consolidation and their deformation analyses.In: Proceedings of the Institution of Civil Engineers Ground Improvement (2014)

DeJong-Hughes, J., Moncrief, J.F., Voethees, W.B., Swan, J.B.: Soil Compaction: Causes, Effects and Control. University of Minnesota (2001)

Federal Emergency Management Agency FEMA. "Geotextiles in Embankment Dams" (2008)

Gruner, L., Floess, C., Micciche, R., Myers, T.: Micropiles on a Mountainside. Foundation and Ground Improvement, Geotechnical Special Publication No. 113, ASCE, Blacksburg VA (2001)

Hayward Baker Geotechnical Construction. "Grouting" (n.d.). www.haywardbaker.com/WhatWeDo/Techniques/Grouting/CementGrouting/default.aspx

Khaldoun, F., Brandon, T.L.: "Soil improvement and Liquefaction mitigation by deep dynamic compaction", Foundation and Ground Improvement, Geotechnical Special Pub. No. 113, ASCE, Blacksburg, VA (2001)

Keller Group UK (n.d.). Ground Improvement. https://www.keller.co.uk/expertise/techniques/

Kowalski, T.E., Starry Jr., D.W.: Modern soil stabilization techniques. In: Annual Conference of the Transportation Association of Canada, Saskatoon, Saskatchewan, October, pp. 14–17 (2007)

Martin, J.R., Mitchell, J.K., Olgun, C.G., Durgunoglu, H.T., Emrem, C.: Preliminary Findings from an Investigation of Improved Ground Performance during the 1999 Turkey Earthquakes. NSF-TUBITAK Turkey Taiwan Earthquakes Grantee Workshop, 24–26 March 2002, Antalya, Turkey (2002)

McCarthy, D.F.: Essential of Soil Mechanics and Foundations: Basic Geotechnics. Prentice Hall, Upper Saddle River (2002)

Moseley, M.P., Kirsch, K.: Ground Improvement. CRC Press (1993)

Nalbantoğlu, Z.: Effectiveness of class C fly ash as an expansive soil stabilizer. Constr. Build. Mater. **18**(6), 377–381 (2004)

Negi, A.S., Faizan, M., Siddharth, D.P., Singh, R.: Soil stabilization using lime. Int. J. Innovat. Res. Sci. Eng. Technol. 448–453 (2013)

Nishimura, S., Takehana, K., Morikawa, Y., Takahashi, H.: Experimental study of stress changes due to compaction grouting. Soils Found. **51**(6), 1037–1049 (2011)

Ramaji, A.E.: A review on the soil stabilization using low-cost methods. J. Appl. Sci. Res. **8**(4), 2193–2196 (2012)

Saye, S.R., Esrig, M.I., Williams, J.L., Pilz, J., Steven, F., Bartlett, S.B.: Lime cement columns for the reconstruction of interstate 15 in Salt Lake City, Utah, Foundation and Ground Improvement, Procs.s, ASCE, Blacksburg VA (2001)

Shakeran, M., Eslami, A., Ahmadpour, M.: Geotechnical Aspects of Explosive Compaction. Shock and Vibration, Hindawi Publishing Corporation (2016)

Sego, D.C., Robertson, P.K., Sasitharan, S., Kllpatrick, B.L., Pillai, V.S.: Ground freezing and sampling of foundation soils at Duncan Dam. Canadian Geotech. J. **31**(6), 939–950 (1994)

Chien, S.-C., Ou, C.-Y., Lo, W.-W.: Electro-osmotic chemical treatment of clay with interbedded sand. Proc. ICE – Geotechn. Eng. **167**(1), 62–71 (2012)

Stapelfeldt, T.: Preloading and vertical drains. Electronic publication. http://www.tkk.fi/Yksikot/Rakennus/Pohja/Preloading_and_vertical_drains.pdf

Numerical Evaluation of Bearing Capacity of Step-Tapered Piles Using P-Y Curves Analysis

Amin Shafaghat[1](✉) and Hadi Khabbaz[2]

[1] School of Civil and Environmental Engineering, University of Technology Sydney (UTS), Ultimo, Australia
Amin.shafaghat@student.uts.edu.au

[2] Deputy Head of School Research, School of Civil and Environmental Engineering, CBIR - Centre for Built Infrastructure Research, University of Technology Sydney (UTS), Ultimo, Australia

Abstract. Step-tapered piles are those with a larger top diameter, and a smaller diameter at lower sections as the body gets slender stepwise from top to toe in one or some steps. This study aims to investigate the behaviour of step-tapered piles having only one step under axial loading condition. Three series of piles embedded in sand are examined numerically using the three-dimensional finite element method. Each set consists of five piles, including one reference straight sided wall pile and four step-tapered piles having the same volume. Different internal friction angles of 33°, 37°, 41° (to represent loose, medium and dense sands, respectively) and corresponding elastic modulus and lateral earth pressure coefficients are considered to observe their effect on the bearing capacity and settlement of piles. The load-displacement diagram of each pile is obtained, and accordingly, the frictional and end bearing resistances are calculated using conventional methods. A MATLAB code is developed to get the numerical data and carry out the calculations. Moreover, the normal and shear stress states, plastic points, and deformations around the step and toe of piles are computed and compared. According to the results, the advantages of step-tapered piles over their counterpart cylindrical ones in terms of bearing capacity and settlement are discussed. Finally, the optimum stepped length of the pile is determined.

1 Introduction

Step-tapered piles can provide bearing capacity through their shaft, base, and the end surface of stepped section. This specific kind of deep foundation has a larger upper diameter (for an assigned pile length) and smaller diameters in the lower parts. Step-tapered piles have a more efficient distribution of material through their body. As the upper sections endure higher forces compared to the lower parts, accumulation of material distribution in the upper part is more efficient to resist against greater forces. Most of the relationships, developed for predicting the behaviour of piles, are associated with the conventional cylindrical piles or prismatic ones, while for step-tapered piles, there are limited analytical solutions. By and large, in the last three decades, the growing tendency

H. Shehata and M. Badr (Eds.): *Advancements in Geotechnical Engineering*, SUCI, pp. 200–212, 2021.
https://doi.org/10.1007/978-3-030-62908-3_15

to investigate the behaviour of tapered and step-tapered piles has made it inevitable to conduct more research in this area. Figure 1 demonstrates some different bored cast piles having various shapes (Bowles, 1996). Referring to Fig. 1 (h), a step-tapered pile is demonstrated schematically. This study aims to investigate the behaviour of these piles installed in sand. The load-displacement diagram of each model is obtained separately, and a detailed comparison is made to quantify the behaviour of piles with different shapes, but having the same volume. By altering the stepped length of the pile, the stress contours developed around the shaft will change due to changes in the stress level and the lateral earth pressures.

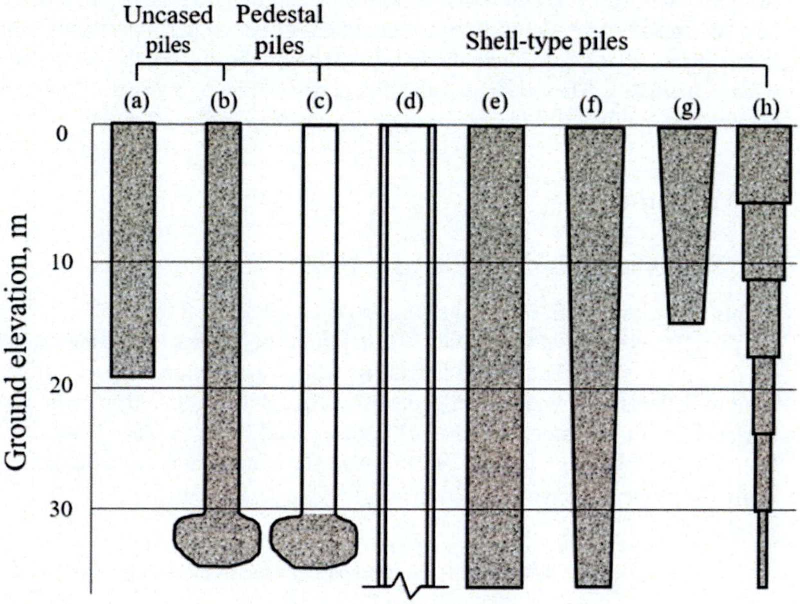

Fig. 1. Some common types of cast-in-place (patented) piles: (a) Commonly used uncased pile; (b) Franki uncased pedestal pile; (c) Franki cased pedestal pile; (d) welded or seamless pipe; (e) Western cased pile; (f) Union or Monotube pile; (g) Raymond standard; (h) Raymond step-tapered pile (after (Bowles, 1996))

2 Literature Review

Nordlund (1963) examined the load-carrying capacity of tapered piles and proposed design diagrams to obtain the coefficient of earth pressure considering the effect of the tapering angle. D'Appolonia and Hribar (1963) investigated the demeanour of step-tapered piles with a corrugated surface taking into account its load transfer mechanism and an analytical method has been proposed accordingly to attain the load transfer curves of these piles. Dutta (1986) performed model pile load tests to study the effect of conical shape of pile with inconstant cross-section on the ultimate compression and uplift capacities and found that the tapering angle was a crucial factor influencing the load capacity of the pile and figured out that cross-section geometry also has a significant effect on the pile load-carrying capacities. Fellenius et al. (2000) analysed the influence of soil set-up and inconstant concrete modulus for tapered instrumented piles in sand.

Bearing capacity of tapered piles is the sum of the tip resistance and the side resistance. The latter comprises two components, (1) skin friction, and (2) vertical component of wedge reaction. Some studies have been performed on a single and a group of step-tapered piles embedded in cemented sand comprising field tests (Ismael, 2001, 2003, 2006 and 2009). Jain et al (2013) examined the comparative demeanour of tapered and straight-sided wall piles driven in loose sands and debated the dependability of N (the bearing capacity factor) and K_h (the coefficient of lateral earth pressure) values. (Hataf & Shafaghat, 2015a, 2015b; Shafaghat, 2013) have investigated the behaviour of tapered piles numerically. They indicated that the performance of a group of tapered piles and the pile group efficiency is a function of the tapering angle.

3 Methodology

3.1 Modelling in PLAXIS 2D

In this study, the width of the cluster is considered to be at least ten times greater than the pile diameter, equal to 6 m, and the depth of the cluster is considered to be at least three times greater than the pile length, equal to 30 m. After creating a soil cluster, the state of underground water level is determined and also some properties of the soil are attributed to this cluster. The level of underground water is placed at a depth of 20 m from the ground. The meshing of the entire soil cluster and pile volume is performed using the properties given in Table 1.

Table 1. Meshing properties used in modelling

Elements type	15-node
Number of soil elements	1381
Number of nodes	11332
The average of elements size (m)	0.57

The meshing of the model is refined at the pile-soil interface. Since piles are considered bored cast in situ, the analysis is performed in three stages; the first stage is the initial condition, which resembles the soil without any structural element in it, the second and the third stages include placing of the pile in the soil and loading on the pile head, respectively. For loading on piles, an estimated load of almost twice the capacity of a pile bearing capacity is used; hence, the obtained load-settlement diagrams will advance enough to apply different methods for bearing capacity calculations.

3.2 Calculating Bearing Capacity of Piles

In order to obtain the bearing capacity of each pile, two loading series are required to be applied. Each pile is initially subjected to a vertical loading that is approximately twice the estimated capacity through analytical solutions. By the completion of the analysis, the load-settlement diagram of the model is obtained. Then, by employing a MATLAB code, written by the authors, and using the double tangential method, the bearing capacity of each pile is calculated. Thereafter, the pertained pile is analysed under the obtained load capacity once again, so that the actual stress distribution proportional to the bearing capacity of each model at the end of the pile or around the pile walls can be observed. Likewise, in order to obtain end the bearing capacity of each pile, the effective stress distribution perpendicular to the end surface elements are acquired and by the use of appropriate mathematical methods the amount of end bearing capacity of the pile is estimated. The frictional resistance of piles can also be calculated by subtracting the two last values from the total bearing capacity. However, that end bearing capacity due to the stepping of piles should be calculated similarity according to the number of steps. According to Fig. 2 and applying Eqs. (1)-(5), the value of end bearing capacity of step-tapered piles can be calculated.

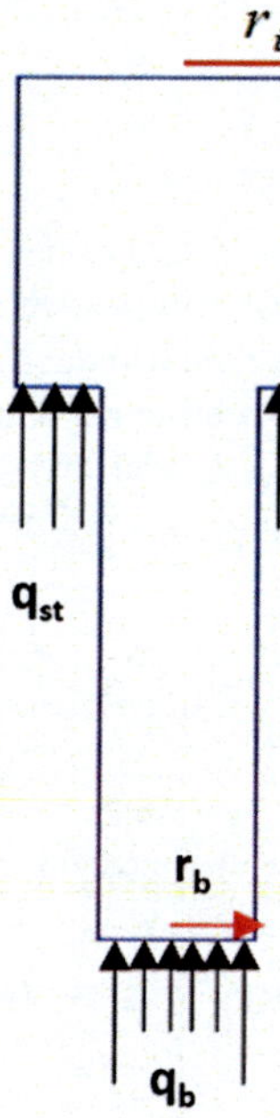

Fig. 2. The schematic shape of a step-tapered pile

The area of the end surface of the pile stepped section is calculated by:

$$A_{st} = \pi r_t^2 - \pi r_b^2 \tag{1}$$

where r_t and r_b are the radii of upper and lower parts of the pile, respectively.

The area of the pile end surface is:

$$A_b = \pi r_b^2 \quad (2)$$

The end bearing capacity due to the stepped section is obtained using:

$$q_{b-st} = q_{st}.\left(\pi r_t^2 - \pi r_b^2\right) \quad (3)$$

where q_{b-st} is the end bearing of stepped section and q_{st} is the effective normal stress on the stepped section.

The end bearing capacity of the pile at its toe is:

$$q_{b-t} = q_b.\left(\pi r_b^2\right) \quad (4)$$

where q_{b-t} is the end bearing of the pile toe and q_b is the effective normal stress on the pile toe.

Hence, the total end bearing capacity is obtained by:

$$q_V = q_{b-st} + q_{b-t} \quad (5)$$

3.3 Model Properties Selection

In this study, three series of piles are analysed separately. Each series consists of five pile models with different geometries. Within each series, all variables of sand are constant, and each five series of piles are analysed under the same loading condition. The first series is examined in loose sand with a friction angle of $\phi = 33°$, a dilation angle of $\psi = 3°$, elastic modulus of 30 MPa and a lateral pressure coefficient of $K_0 = 0.45$. The key properties of 3 series of soils, implemented in the numerical analysis, are summarised in Table 2.

Table 2. Different considered conditions for modelling in sand

Set 3 (dense sand)	Set 2 (medium sand)	Set 1 (loose sand)
41° = ϕ	37° = ϕ	33° = ϕ
E = 60 MPa	E = 45 MPa	E = 30 MPa
11° = ψ	7° = ψ	3° = ψ
0.33 = K	0.40 = K	0.45 = K

Three series of pile models in sand with different ϕ, ψ, elastic modulus and different at rest lateral earth pressure coefficients are analysed and compared. Then comparison is made for models within a series and also for series with each other. To illustrate the point, consider the first series of piles, which consisted of five piles, including a cylindrical pile and four step-tapered piles that can be compared to each other. Pile models are selected and made in a way to be practically applicable in real scale. Stepped models with a cylindrical reference model are possible to be made in two ways. The first method is to maintain the upper cross-section of pile constant and reduce the lower cross-section of the pile. The volume of step-tapered piles constructed with this method in comparison with the cylindrical reference pile is reduced. Therefore, this method is out of the scope of this research. The second method is to consider a cylindrical reference pile and change the upper and lower section of pile simultaneously. Hence, the smaller lower cross-section and the larger upper cross-section of pile are made and by a change in the length of the upper part of the pile, different models of piles having the same volume are made. Since the purpose is a comparison of the bearing capacity and the settlement of step-tapered piles with their counterpart cylindrical ones having the same volume, the research is focused on the second method. The length of the cylindrical reference model is considered 10 m, and its diameter is selected 0.6 m. Hence, by changing the length and the diameter of the upper part of the pile, different pile models are created. The properties of different step-tapered models are presented in Table 3.

Table 3. Properties of step-tapered piles having the same volume and cylindrical reference piles

Pile	$L' = {}^{h_i}/_{h} * 100$	r_t	r_b	L_t	L_b	H (m)	V (m^3)
C	100	0.3	0.3	10	–	10	2.8
ST1	20	0.45	0.25	2	8	10	2.8
ST2	30	0.4	0.25	3	7	10	2.8
ST3	40	0.36	0.25	4	6	10	2.8
ST4	50	0.34	0.25	5	5	10	2.8

In Table 3, h_i is the length of the upper part of the step-tapered pile, h is the total length of the pile and $h_i/h\ ^*100$ is the ratio of the length of the upper part to the total length of the pile, or in other words it is the percentage of the total length of the pile with a greater diameter. L_t and L_b are the height of the upper and lower sections of the stepped section, r_t and r_b are the radii of the upper and lower parts of the pile respectively and V is the total volume of the pile. In this study in order to make stepped models of the same volume, the diameter of the lower part of the pile for all models is considered constant and equal to 0.5 m and the diameter of the upper part of the pile changes. Therefore, there can be different pile models as shown in Fig. 3.

Pile models of the study are made of concrete material and a linear elastic model has been designated for piles. The assigned properties of concrete, used in simulations, are given in Table 4.

Since soil-facing mechanical interactions play a significant role in obtaining the pile bearing capacity, numerical analysis of pile foundations entails the use of interface elements between different component materials to analyse the soil-structure interactions

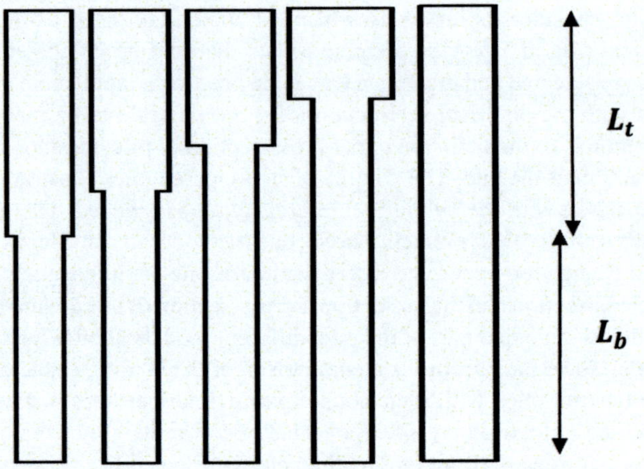

Fig. 3. The schematic shape of step-tapered along with their cylindrical reference pile

Table 4. Properties of used concrete in modelling

Concrete	Material
Linear elastic	Material model
Non-porous	Drainage type
27	$\gamma(\frac{kN}{m^3})$
$31*10^6$	$E(\frac{kN}{m^2})$
0.1	ν

and to capture the transfer of shear and normal stresses through these discontinuities. When using finite element method in PLAXIS software, soil-structure interactions can be modelled using "zero-thickness" interface elements between the soil and pile body components. These elements use a stiffness and strength reduction factor that is applied to the soil strength parameters in the interface zone. In order to consider the interaction effect between the soil and concrete interface, the strength value in this study is considered as $R_{inter} = 0.7$. Figure 4 illustrates the step-tapered pile (ST3) model embedded in sand along with the point load on pile head and the pile-soil interface.

3.4 The Properties of Used Soil

Soil selection for the purpose of modelling has been made so that the results can be verified with a practical case. For this purpose, the results of three common types of sandy soil (loose, medium, and dense) were evaluated and compared. In sandy soil, more focus is on the internal friction angle, dilation angle and the lateral pressure coefficient of the soil. The soil properties are selected and shown in Table 5. As mentioned, for the

Fig. 4. Step-tapered pile model in Plaxis 2D along with the pile-soil interface

purpose of sensitivity analysis, the internal friction angle (ϕ), the dilation angle (ψ), and the lateral pressure coefficient (K) are considered as three variables.

Table 5. Properties of used sand in modelling

$\gamma(\frac{kN}{m^3})$	γ_{sat} $(\frac{kN}{m^3})$	Material model	E (MPa)	ϕ′ (degree)	C′ (kPa)	ψ	K_o	υ	R_{INTER}
17	20	Hardening soil (Drained)	30 45 60	33° 37° 41°	0	3° 7° 11°	0.45 0.4 0.33	0.2	0.7

4 Results and Discussion

The load-displacement diagrams of models are obtained and finally to determine the effectiveness of step-tapered piles a comparison is made.

4.1 Results of Bearing Capacity and Settlement

After conducting several numerical analyses, the p-y curve of each straight-sided wall pile and step-tapered pile obtained. Figure 5 illustrates the load-settlement curves of piles in sand with an internal friction angle of 33°.

End bearing capacity of each pile is assessed using numerical integration of effective normal stress on the pile toe and stepped toe as well. Then frictional resistance of piles

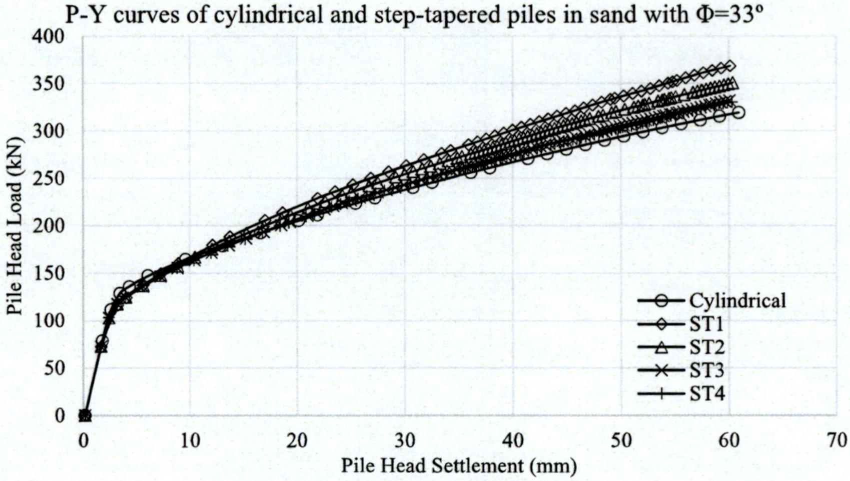

Fig. 5. The P-Y curves of cylindrical reference pile and step-tapered piles in sand with $\varphi = 33^{\circ}$

is calculated by subtracting the end bearing component from total resistance of pile. It should be noted that the bearing capacity of step-tapered piles is the sum of the three values: 1- The friction values of the body; 2-End bearing of the pile toe; 3-End bearing of the stepped part of the pile.

4.2 Stress States and Plastic Points Around Piles

According to Figs. 6 and 7, for cylindrical piles under axial load, the stress direction and plastic points are beneath the pile toe. A straight-sided wall pile has a larger toe cross-sectional area than that of its counterpart same volume step-tapered pile. Hence, the end bearing capacity provided by cylindrical piles is only at the base and is more than that of step-tapered piles. However, for step-tapered piles, the end resistances is achieved by mobilising the soil beneath the toe and the stepped part as well. Indeed, due to the soil densification beneath the stepped part, the shaft resistance boosts. As can be seen in Fig. 6, the soil beneath the stepped part densifies the soil adjacent to the pile. Therefore, the lateral earth pressure coefficient of the soil increases, and consequently, the effective lateral pressure and shaft resistance of pile are boosted.

As Figs. 6 and 7 suggest, the stepped part of the step-tapered piles can provide the shaft with higher resistances. The distribution of plastic points illustrates that cylindrical piles are not using all the potential shaft capacity as a larger area around the shaft is still in the elastic zone. Hence, existing a step on the pile body can increase the pile resistance by providing more capacity through more hardening and cap points. Tables 6, 7 and 8 illustrate some differences between the bearing capacity of step-tapered piles and their counterpart same volume cylindrical piles for three different series of models.

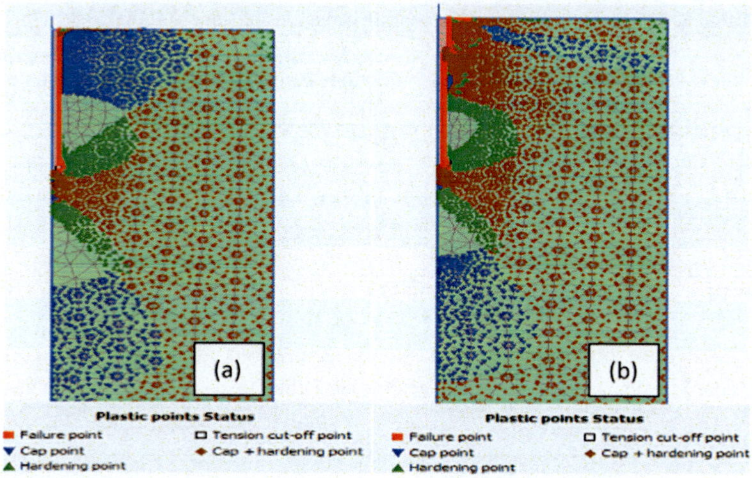

Fig. 6. Distribution of plastic points around a) Cylindrical pile b) Step-tapered ST1

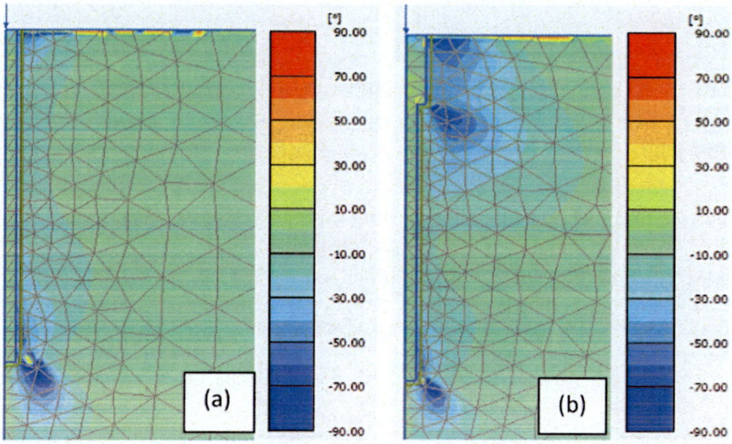

Fig. 7. Principal stress directions around a) Cylindrical pile b) Step-tapered ST1

4.3 The State of Stress on the End Surface of Piles (Base)

The state of shear stress on the pile shaft is indicated in Fig. 8 for two cylindrical and step-tapered (ST4) piles. According to the figures below, it can be perceived that for cylindrical piles, the shear stress state is relatively high at the pile toe, while for step-tapered piles the shear stress state is higher at both the stepped part and pile toe. As can be seen, providing higher shear stresses beneath the stepped part can increase the shaft capacity of the pile. Hence, this proves that step-tapered piles have the potential to be used in granular soil or even layered soil.

Table 6. Comparison between the bearing capacities of same volume piles in loose sand (Set 1)

Piles	End bearing capacity (kN)	Friction bearing capacity (kN)	Total bearing capacity (kN)
C	170	54	224
ST1	192	69	261
ST2	181	57	238
ST3	176	56	232
ST4	174	55	229

Table 7. Comparison between the bearing capacities of same volume piles in medium sand (Set 2)

Piles	End bearing capacity (kN)	Friction bearing capacity (kN)	Total bearing capacity (kN)
C	457	152	609
ST1	500	203	703
ST2	520	173	693
ST3	495	165	660
ST4	465	155	620

Table 8. Comparison between the bearing capacities of same volume piles in dense sand (Set 3)

Piles	End bearing capacity (kN)	Friction bearing capacity (kN)	Total bearing capacity (kN)
C	716	306	1022
ST1	803	373	1176
ST2	769	329	1098
ST3	740	317	1057
ST4	731	313	1044

5 Conclusions

In the current study, the bearing capacity and settlement of step-tapered piles under axial load are analysed and compared with their counterpart same volume cylindrical piles.

The maximum bearing capacity is provided from the stepped pile which has the shortest step (ST1) with the bearing capacity ratio of around 1.15 with respect to its counterpart same volume cylindrical pile.

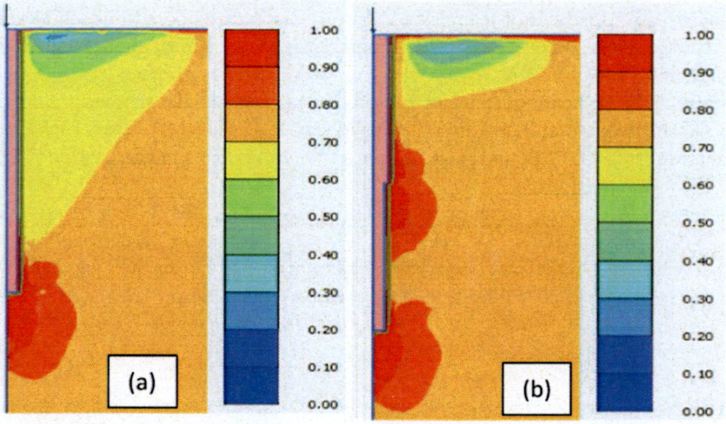

Fig. 8. Relative shear stress distribution on the pile shaft a) Cylindrical pile b) Step-tapered pile

For step-tapered piles placed in sand, due to the soil densification and compaction beneath the stepped section lateral earth pressure coefficient of the soil increases and consequently the shaft resistance and finally total bearing capacity increase.

Step-tapered piles provide more end bearing capacity through the base and stepped section, thus these piles can be an alternative to conventional straight-sided wall piles.

Step-tapered piles might be an efficient deep foundations in layered soil where the stepped section can be design to be located on top of each layer to increase the vertical effective stress adjacent to the pile. Hence, where the pile body faces a cohesive soil, the stepped section can increase the end resistance and where the body interacts with a non-cohesive soil, the shaft resistance can provide capacity.

References

Bowles, L.E.: Foundation Analysis and Design. McGraw-Hill, New York (1996)

D'Appolonia, E., Hribar, J.: Load transfer in a step-taper pile. J. Soil Mech. Found. Div. **89**(6), 57–80 (1963)

Dutta, S.: Influence of surface taper and shape of pile on ultimate load and uplift capacities. Indian Geotech. J. **16**, 167–180 (1986)

Fellenius, B.H., Brusey, W.G., Pepe, F.: Soil set-up, variable concrete modulus, and residual load for tapered instrumented piles in sand. In: Performance Confirmation of Constructed Geotechnical Facilities, pp. 98–114 (2000)

Hataf, N., Shafaghat, A.: Numerical comparison of bearing capacity of tapered pile groups using 3D FEM. Geomech. Eng. **9**(5), 547–567 (2015a)

Hataf, N., Shafaghat, A.: Optimizing the bearing capacity of tapered piles in realistic scale using 3D finite element method. Geotech. Geol. Eng. **33**(6), 1465–1473 (2015b)

Ismael, N.: Load tests on straight and step tapered bored piles in weakly cemented sand. In: Nabil, F.I. (ed.) Field Measurements in Geomechanics, pp. 142–149. CRC Press, Boca Raton (2003)

Ismael, N.F.: Axial load tests on bored piles and pile groups in cemented sands. J. Geotech. Geoenvironmental Eng. **127**(9), 766–773 (2001)

Ismael, N.F.: Analysis of lateral load tests on step tapered bored piles in calcareous sands. In: GeoCongress 2006: Geotechnical Engineering in the Information Technology Age, pp. 1–6 (2006)

Ismael, N.F.: Behavior of step tapered bored piles in sand under static lateral loading. J. Geotech. Geoenvironmental Eng. **136**(5), 669–676 (2009)

Jain, M., Rastogi, P., Bhandari, R.: Comparative behavior of tapered and uniform diameter piles in loose sands (2013)

Nordlund, R.: Bearing capacity of piles in cohesionless soils. J. Soil Mech. Found. Div. **89**(3), 1–36 (1963)

Shafaghat, A.: Numerical comparison of bearing capacity of tapered and cylindrical pile groups with three-dimensional finite element method. (MSc). Shiraz University, Iran (2013)

Using Traditional and Advanced Soil Improvement Techniques for Swelling Soil Heave Mitigation

G. E. Abdelrahman, Y. G. Youssef, and A. E. Abdeltawab(✉)

Soil Mechanics and Foundation Engineering, Civil Engineering Department, Faculty of Engineering, Fayoum University, Fayoum, Egypt
aea12@fayoum.edu.eg

Abstract. Swelling soil is one of the most common problematic soils. The swelling phenomenon can cause severe damage to the overlying lightly loaded structures. Many alternatives have been proposed to mitigate the effects of expansive soil on civil foundations such as pre-wetting, sub-drainage, hydraulic barriers, chemical stabilization, and mechanical treatments. Sand replacement is considered to be one of the most effective and practical techniques that can be utilized for the construction of safe foundations. The sand cushion is used in this study as a traditional solution to reduce the heave of the expansive soil. On the other hand, the sand cushion reinforced with triaxial geogrid is used as an advanced solution. An experimental study using a large-scale box model is used to investigate the effect of sand cushion on reduces the heave of the expansive soil. Besides, a parametric study was performed to assess the effect of sand cushion thickness and lateral extension on the heave of the footing. Also, the effect of reinforcing sand cushion with triaxial geogrid was taken into consideration and studied the importance of using it in reducing of the swelling soil heave. The results demonstrated that the use of sand cushion reduces the heave of the footing by increasing the thickness of the sand cushion. Moreover, the reduction rate of heave decreases for sand cushion thickness greater than two-thirds of the footing width. Thus, the optimum thickness of the sand cushion is 2/3 footing width. The footing heave decreases with increasing the lateral extension, *Le*, of sand cushion, and the rate of reduction of heave becomes negligible when using lateral extension greater than twice thickness of sand cushion. Furthermore, it was concluded that using triaxial geogrid with sand cushion as a reinforcement element leads to reduce heave of the swelling soil due to the tension developed in geogrid. In addition, The maximum reduction percentage of the footing heave is equal to 53.77% when using sand cushion thickness is equal to the width of footing but, the minimum reduction percentage of heave is equal to 35.39% when using sand cushion thickness of 1/3B.

1 Introduction

Many buildings, pavements, airport runways, retaining walls, dams, and bridges are distressed by swelling soils. These natural high plasticity soils typically contain clay

H. Shehata and M. Badr (Eds.): *Advancements in Geotechnical Engineering*, SUCI, pp. 213–224, 2021.
https://doi.org/10.1007/978-3-030-62908-3_16

mineral montmorillonite that exhibits high swelling with the increase of water content (Chen, 1988). These types of soils are mostly found in arid and semi-arid climatic regions. Many locations in Egypt are well-known by the presence of swelling soils such as; Nasr – City, Cairo – Suez Road, El – Fayoum City, Kom Ombo City, Aswan City, New Valley, El – Sadat City, and Sohag – Safaga highway (Elbeih & Soliman, 2015). With these soils experiencing volume changes, shallow foundation may exhibit surficial distress due to the swell-shrink behavior which leads to cracks in these structures.

Sand replacement is considered to be one of the most effective and practical techniques that can be utilized for the construction of safe foundations (Awad, 2005). This method is used because of its many advantages such as the sand replacement can be compacted to a higher density, which leads to higher strength, does not require specialized equipment, and it requires no soil additives. However, it does have some disadvantages such as the high volume of excavation may be costly, even after excavation, the underlying and adjacent expansive soils often must be protected by horizontal or vertical membranes.

There is a lot of researches that have examined and investigated the use of sand cushion to control the heave of the swelling soil such as Ahmed, 1985 concluded that providing isolated footing with a sand cushion as a good practice to minimize the swelling pressure as it is possible to reduce the swelling pressure by 25% when using sand replacements with a thickness equal to one-fifth of the footing width.

El-Nahhas et al., 1998 studied the effects of the magnitude of active depth, *Ha*, and the thickness of soil replacement on the heave of swelling soil. The results of tests show if 40% of the maximum active thickness of an expansive layer is removed and replaced with compacted non-expansive soil, the maximum heave would be only 54% of that anticipated with no soil replacement. Awad, 1993 concluded that the reduction of final footing heave in case of treated expansive soil using sand replacement is 30% and 52% when the sand thickness is half and one of the footing widths, respectively.

Youssef et al., 2010 investigated the effect of sand cushion parameters on the swelling behavior of expansive soils subjected to changes in soil suction profiles due to climate variations by using a numerical model. The analysis presented provides considerable insight into the effect of the sand cushion as a mitigation method on the behavior expansive soils. Conclusions may be summarized as follow: sand cushion depth has a significant effect on decreasing footing heave and the lateral extension of the sand cushion has a significant effect on the heave of footing. However, increasing lateral extension more than twice the depth of sand cushion is insignificant on footing heave.

Daifalla and Witt, 2014 carried out the swelling tests on a mixture of bentonite-silt reinforced with a layer of geogrid placed horizontally at the mid-height sample. The test results showed that the geogrid reduced the swelling potential. Niwas et al., 2017 use geocomposite (geogrid and geotextile were switched together) to control the expansion of swelling clays with bentonite and after performing all the tests using conventional one-dimensional oedometer apparatus, it was concluded that for any geosynthetic material, the one which is confined found to control the swelling 25 to 35% higher than the same material without confinement.

Treatment of expansive soil beneath isolated footing using sand replacement is considered one of the most effective, and practical techniques that can be utilized for the construction of safe shallow foundations. The objective of this research is to investigate the effect of sand cushion thickness and lateral extension on the heave of the footing models resting on expansive soil. Also, the effect of reinforcing sand cushion with triaxial geogrid was taken into consideration and studied the importance of using it in reducing of heave of swelling soil.

2 Materials

2.1 Sand Cushion

Fine to medium sand soil is used as layers of sand cushion. Laboratory tests such as sieve analysis test, specific gravity test, compaction test, and direct shear test were conducted to assess sand properties. Table 1 shows the sand properties.

Table 1. Sand cushion properties

Properties	Values
Bulk unit weight (KN/m^3)	14.8
Specific gravity, *Gs*	2.65
Fine (%)	1.16
Coefficient of uniformity, C_U	3.33
Coefficient of curvature, C_C	1.57
Unified soil classification	SP
Optimum moisture content, *O.M.C*, (%)	11
Max-dry density, *MDD*, (KN/m^3)	18.6
Internal friction angle, φ	34

2.2 Expansive Soil

In this study, the swelling soil is remolded from 40 percent bentonite and 60 percent kaolinite clay. The laboratory tests carried out on expansive soil were Atterberg limits, hydrometer test, compaction test, free swell test, specific gravity, and swelling pressure test. The soil used is classified as high plasticity clay (CH) according to Unified Soil Classification System. The swelling soil properties are shown in Table 2.

Table 2. Swelling soil properties

Properties	Values
Specific gravity, *Gs*	2.57
Silt content (%)	36
Clay content (%)	64
Liquid limit, *L.L*, (%)	134
Plastic limit, *P.L*, (%)	34
Plasticity index, *P.I*, (%)	100
Shrinkage limit, *Sh.L* (%)	12
Optimum moisture content, *O.M.C*, (%)	21
Max-dry density, *MDD*, (KN/m^3)	16
Free swell, *F.S*, (%)	350
Swelling pressure, *Ps*, (kpa)	260

2.3 Geogrid

Triaxial geogrid was used for reinforcing the sand cushion. Figure 1 shows triaxial geogrid which was used in the experimental study. This material is composed of polypropylene resin. The properties of the geogrid as supplied by the manufacturer (Tensar) are given in Table 3.

Sample in the test apparatus **Geogrid details**

Fig. 1. Illustration of triaxial geogrid

Table 3. The triaxial geogrid properties (TX160)

Product characteristic		Units	Declared value	Tolerance
Polymer		———	Polypropylene	———
Radial secant stiffness	at 0.5% strain	KN/m	390	−75
	at 2% strain	KN/m	290	−65
	ratio	———	0.8	−0.15
Junction efficiency		%	100	−10
Hexagon pitch		mm	80	±4
Weight		Kg/m^2	0.220	−0.035

3 Experimental Work

A large-scale box model is built to measure the heave of footing due to expansive soil. The dimensions of experimental model are shown in Fig. 2. Also, a photo of the model at the laboratory is shown in Fig. 3. A square steel block with dimensions of 100 × 100 × 10 mm was used to simulate a rigid footing condition.

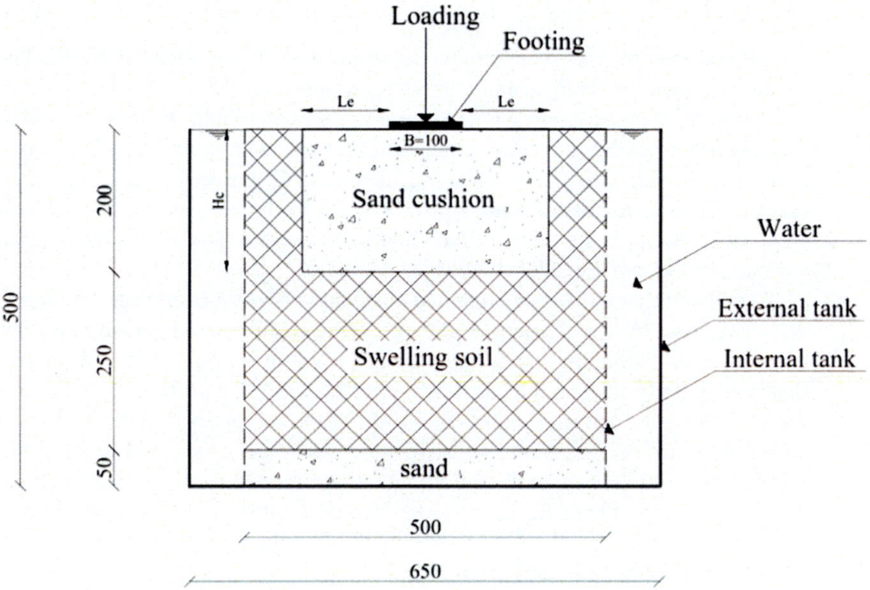

Fig. 2. Experimental model sketch (dimensions are in mm)

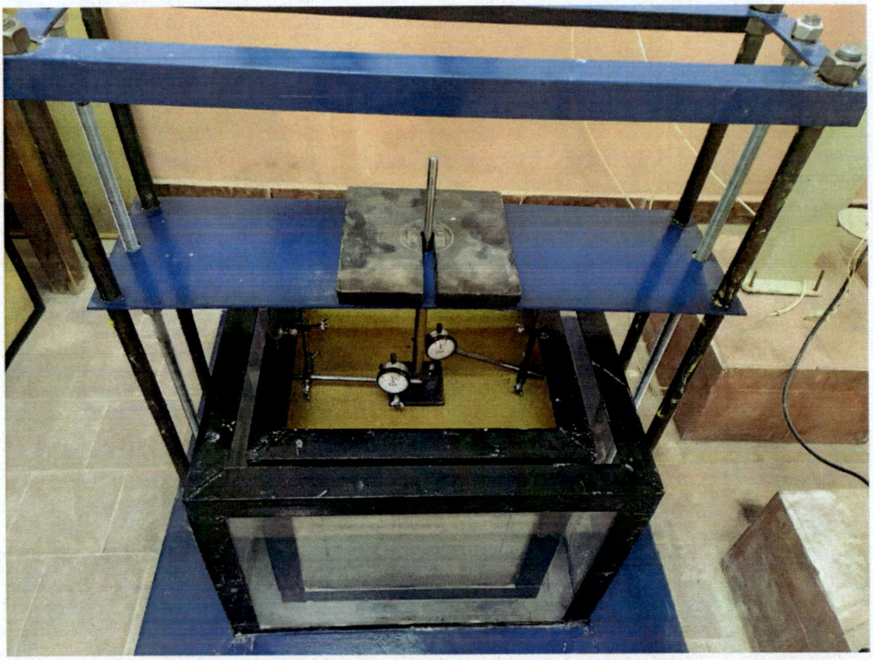

Fig. 3. The large-scale box model at the laboratory

Several experiments were conducted to investigate the effect of the thickness and lateral extensions of sand cushion on the footing heave. The soil is placed in the model and compacted on layers to the maximum dry density. Then the footing is placed and loading begins with the filling of the external tank with water. The heave of footing is measured using two dial gauges until insignificant movement is observed. The soil specimen begins to absorb water from the holes in the perforated plate. The absorbed water enters between the clay platelets causes soil to swell. The footing heave is measured for this method and the axial swell is calculated.

Sand bed thickness was maintained constant as 50 mm in all tests, and the swelling soil thickness, *Hs*, was constant as 250 mm. Sand cushion and expansive clay have been compacted to maximum dry density. Weight loads of 50 kg (50 kPa) were applied on the footing. Sand cushion thickness, *Hc,* of 0, 1/3*B*, 2/3*B*, and *B* are used when studying the effect of sand cushion thickness on the heave of footing where, B is the footing width. Also, lateral extension, *Le*, of the sand cushion is used equal to *Hc*, 1.5*Hc*, 2*Hc*, 2.5*Hc*, 3*Hc*, and 4*Hc* to assess the impact of lateral extension of sand cushion on the heave of footing, where *Hc* is the sand cushion thickness. Besides, the swelling soil, improved using sand cushion reinforced with triaxial geogrid with a different thickness which varied from 1/3B, 2/3B, and B.

4 Test Results and Analysis

The aim of these tests is to study the effect of thickness and lateral extension of sand cushion layers resting on expansive soil on the footing heave. When the test begins and water enters the internal model from its sides, the space between the external tank and the perforated plate of the internal tank is filled with water. In the next sections, each test was performed, finding the footing heave, and then analyzing these results carefully.

4.1 Effect of Sand Cushion Thickness

The effect of variation of sand cushion thickness, *Hc*, on the heave of footing was studied. It is varied as follows: 0, 1/3*B*, 2/3*B*, and *B*, where *B* is the width of the footing. The relationship between time and axial swell, *A.S*, for swelling soil using a sand cushion is shown in Fig. 4. Axial swell, *A.S*, is equal to the ratio of footing heave to swelling soil thickness (*ΔH/Hs*). Axial swell is 6.04%, 4.308%, 3.468%, and 3.10% for the sand cushion thickness are equal to 0, 1/3*B*, 2/3*B*, and *B*, respectively.

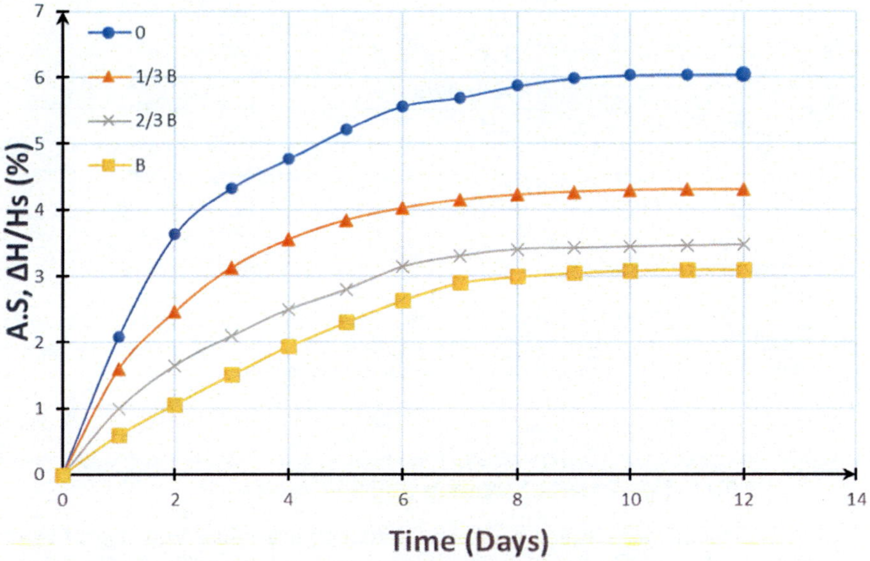

Fig. 4. Time-axial swell curves for variation thickness of the sand cushion

It can be found that the footing heave decreases with the thickness of the sand cushion increasing. Also, it's obvious that within the first six days of testing the rate of swelling is much higher than it started to decrease. By the end of the eleventh day, the rate began reaching a negligible value. The maximum reduction percentage of the footing heave was 48.68% when used sand cushion thickness is equal to the width of footing. Moreover, this reduction percentage reduces with decreasing sand cushion thickness to the minimum value of 28.67% at a thickness of 1/3B. Also, it was found that the heave reduction rate decreases with increasing thickness of the sand cushion.

There are many previous studies that deal with the use of sand cushion to reduce footing heave. Table 4 shows a comparison of the percentage of reduction in heave between this research and previous researches. From this table, it can be said that with the different types of swelling soil used and the way tests are conducted in these studies, however, the reduction percentage in the heave is very similar in all these studies. For example, in the case of using sand as a replacement soil over swelling soil with a thickness equal to the width of the footing, the percentage of reduction in heave was equal to 48.68, 52, and 54 for this study, (Awad, 1993), and (Fathalla El-Nahhas et al., 1998), respectively. Figure 4 illustrates the obtained results for different sand cushion thickness versus the obtained heave of footing. These data showed the same trend obtained by, (El-Dean et al., 2016)

Table 4. Comparison with previous research

References	Sand cushion thickness, *Hc*	Reduction of footing heave (%)
This study	1/3 *B*	28.67
	2/3 *B*	42.58
	B = 0.40 *Ha*	48.68
(Fathalla El-Nahhas et al., 1998)	0.40 *Ha*	54
(Awad, 1993)	0.50 *B*	30
	B	52
(Abouleid and Reyad, 1985)	0.38 *B*	31%
(Ahmed, 1985)	0.20 *B*	25%

Note: *B* = width of footing, *Ha* = active depth of swelling soil.

4.2 Effect of the Lateral Extension of Sand Cushion

The effect of sand cushion width on the footing heave has been studied by changing the lateral extension, *Le*, of sand cushion to be *Hc*, 1.5*Hc*, 2*Hc*, 2.5*Hc*, 3*Hc*, and 4*Hc*. The lateral extension, *Le*, is defined as the extending distance of sand cushion from the boundaries of the footing in each side as shown in Fig. 2. The sand cushion thickness, *Hc*, remains constant and is equal to 0.5B. Figure 5 shows the time-axial swell curves for the variation of lateral extension, *Le*, of a sand cushion. In addition, the values of the maximum *A.S* for each case are illustrated in Fig. 6.

Based on Fig. 5 and Fig. 6, the lateral extension, *Le*, of the sand cushion has a significant effect on the heave of footing. Also, it can be observed that the heave of the footing decreases with increasing the lateral extension, *Le*, of a sand cushion. The maximum axial swell value is equal to 5.426% for lateral extension is equal to the sand

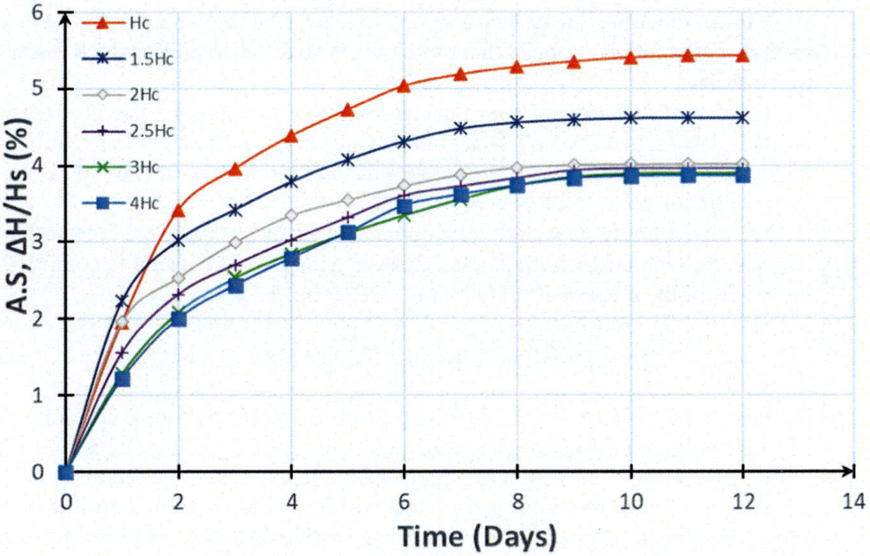

Fig. 5. Time-axial swell curves for variation of lateral extension of sand cushion

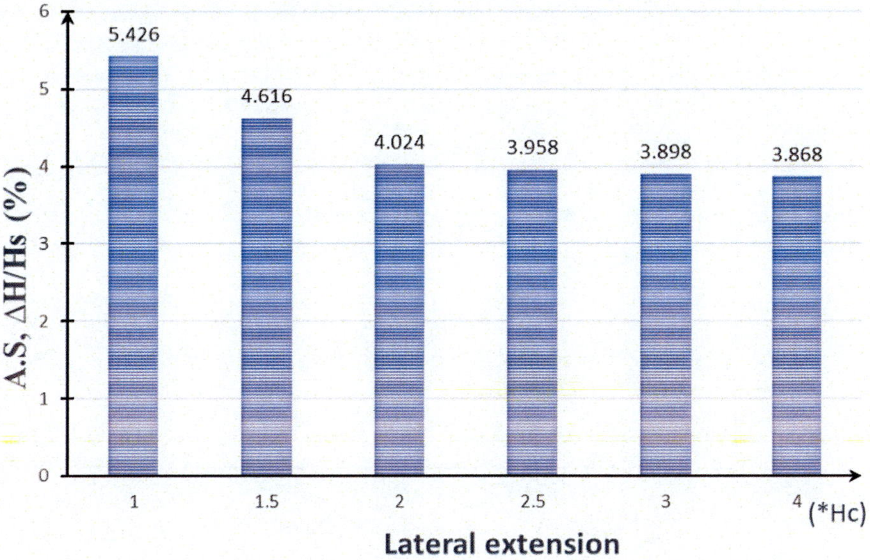

Fig. 6. Maximum axial swell for variation of sand cushion lateral extension

cushion thickness, while this value reduced to reach the minimum value is 3.868% with increasing *Le* to 4*Hc*.

It is clear that the rate of heave started to reach a negligible value by the end of the tenth day. Also, it was noticed that more than 90% of the total heave was reached after the sixth day.

The rate of reduction of heave decreases with increasing the lateral extension, *Le*, of sand cushion, and it becomes negligible when using *Le* greater than 2*Hc*. Therefore, increasing lateral extension, *Le*, of sand cushion more than twice the thickness of the sand cushion is not recommended.

The current study showed that the optimal lateral extension of the sand cushion layer is twice the sand cushion thickness. This experimental conclusion is compatible with the results of the numerical study by (Youssef et al., 2010).

4.3 Effect of the Sand Cushion Reinforced with Triaxial Geogrid

The results of the tests which show the effect of using sand cushion reinforced with triaxial geogrid on the heave of footing due to the swelling soil are presented. Geogrid reinforcement is placed within the sand cushion in a horizontal orientation in the mid-height of sand layer. The sand cushion thickness is as follows: 1/3*B*, 2/3*B*, and B. Figure 7 shows the time-axial swell curves of footing when using sand cushion reinforced with triaxial geogrid.

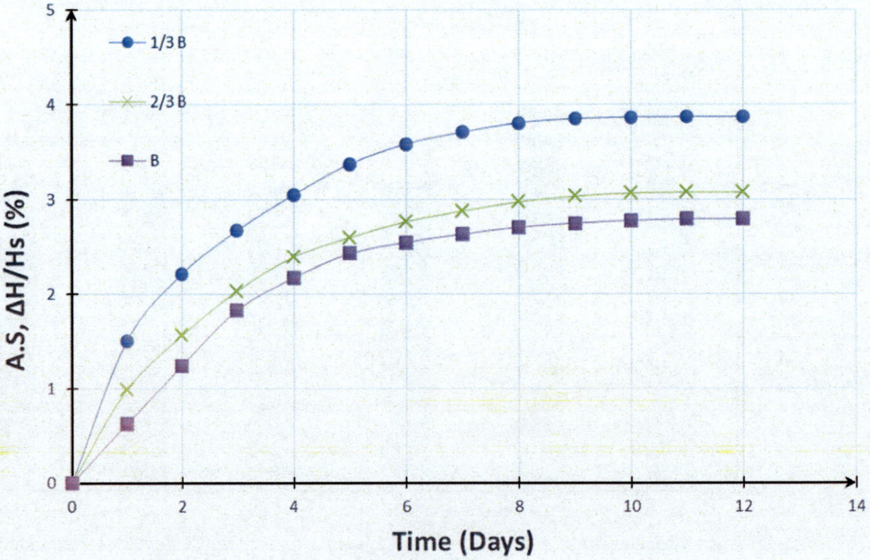

Fig. 7. Time-axial swell curves for variation thickness of the sand cushion reinforced with triaxial geogrid

Based on Fig. 7, it was noticed that the heave of footing due to the use of sand cushion reinforced with geogrid decreases with increasing the thickness of the sand cushion. Besides, it was found that the maximum axial swell, A.S, is equal to 3.324%

which occurs when using the sand cushion thickness of $1/3B$, but the minimum axial swell is 2.578% when used sand cushion thickness is equal to the width of footing. As well, the maximum reduction percentage of heave is equal to 53.77%, but the minimum reduction percentage is 35.93%. Thus, the use of triaxial geogrid in sand cushion reinforcement has a major impact on controlling the heave of the swelling soil mainly due to the effect of the tension developed in geogrid.

5 Conclusions

From analysis of results, the following conclusion may be drawn as:

1. Heave of footing decreases with increasing the thickness of sand cushion, Hc.
2. The rate of decrease in the heave decreases with increasing sand cushion thickness.
3. The maximum axial swell ($\Delta H/Hs$) is 6.04% for swelling soil without sand cushion, while it reduced when using a sand cushion to reach the minimum axial swell is 3.10%.
4. The lateral extension, Le, of sand cushion has a significant effect on the heave of footing.
5. The rate of reduction of heave decreases with increasing the lateral extension, Le, of sand cushion, but it becomes negligible when using lateral extension greater than $2Hc$.
6. The use of triaxial geogrid in sand cushion reinforcement has a major impact on controlling the heave of the swelling soil mainly due to the effect of the tension developed in geogrid.
7. The maximum benefit of geogrid can be obtained if using the sand cushion layer with a small thickness.
8. For the use of sand cushion reinforced with triaxial geogrid, the maximum reduction percentage of the heave is equal to 53.77%. But, the minimum percentage is equal to 35.39%.

6 Recommendation

Based on findings from this research study, the following recommendations are made for future research efforts: carrying out numerous tests on sand reinforced with fixed end condition geogrid (biaxial and triaxial), and investigate its effect on the heave of footing due to the swelling soil by studying many variables such as the location and width of geogrid. Also, the use of other types of geosynthetic in reinforcing the sand cushion such as geonets, geomembranes, and geocomposites.

References

Abouleid, A.F., Reyad, M.M.: Study of the behaviour of isolated footings on expansive soils. In: Proceeding of The Eleventh International Conference on Soil Mechanics and Foundation Engineering, 12–16 August 1985, Publication of: Balkema (AA), San Francisco (1985)

Ahmed, A.A.: Study on the behavior of some Egyptian expansive soils. Thesis, Faculty of Engineering, Ain Shams University (1985)

Awad.: Effect of expansive soil on imposed stress of footing and supper structure. Thesis, Faculty of Engineering. Ain Shams University (1993)

Awad, T.: (2005). Properties of expansive soil treated with sand-lime cushion. In: Tenth International Colloquium on Structural and Geotechnical Engineering, Faculty of Engineering, Ain Shams University, vol. 16, pp. 1–12

Chen, F.H.: Foundations on Expansive Soils. In: Found on Expansive Soils (1988)

Daifalla, S., Witt, K.J.: Geogrid effects on the swelling characteristics of geotextile-encased treated & untreated expansive soils. In: 10th International Conference on Geosynthetics, ICG 2014, 1991. (2014)

El-Dean, M.R.G., Fouad, M.T., Bahr, M.A.: Characteristics of some unsaturated soils based on field and laboratory tests. Thesis, Eng. Mohamed Ragab Gamal El-Dean Under the Supervision of Dr. Mohey Elmashaad, Faculty of Engineering Civil Engineering Department (2016)

El-Nahhas, F., Awad, T., Abd-Moatty, N.: Prediction of Ground heave due to the swelling of multi-layered soil deposits. In: 8th International Colloquium on Structural and Geotech. Engineering, Faculty of Engineering, Ain Shams University (1998)

Elbeih, S.F., Soliman, N.M.A.: An approach to locate and map swelling soils around Sohag-Safaga road, Eastern Desert, Egypt using remote sensing techniques for urban development. Egypt. J. Remote Sens. Space Sci. **18**(1), S31–S41 (2015). https://doi.org/10.1016/j.ejrs.2015.07.002

Fathalla E., Awad, T., Abdel-Moatty, N.: Prediction of Ground Heave Due to Swelling of Multi-Layered Soil Deposits. In: Eighth International Colloquium on Structural and Geotechnical Engineering 15 – 17 December, L99g Ain Shams University Cairo, Egypt (1998)

Niwas, R., Laddha, A., Purohit, D.G.M.: Experimental study on control of swelling pressure of expansive soils using bentonite and one layer of geosynthetics. 14(5), 9–15 (2017) https://doi.org/10.9790/1684-1405040915

Youssef., Darrag, A.A., Elkady, T.Y.: Numerical Modeling of Expansive Soils Using Numerical Modeling of Expansive Soils Using. Ph.D. (2010)

Author Index

H. Shehata and M. Badr (Eds.): *Advancements in Geotechnical Engineering*, SUCI, pp. 225–226, 2021.
https://doi.org/10.1007/978-3-030-62908-3